Biophilic Cities for an Urban Century

Robert McDonald • Timothy Beatley

Biophilic Cities for an Urban Century

Why nature is essential for the success of cities

palgrave
macmillan

Robert McDonald
Arlington, VA, USA

Timothy Beatley
Charlottesville, VA, USA

ISBN 978-3-030-51664-2 ISBN 978-3-030-51665-9 (eBook)
https://doi.org/10.1007/978-3-030-51665-9

This Palgrave Pivot imprint is published by the registered company Springer Nature Switzerland AG.
The registered company address is: Gewerbestrasse 11, 6330 Cham, Switzerland

Acknowledgments

The authors thank their respective institutions for their support while working on this book, as well as all the dedicated people working on nature-based solutions and biophilic design in cities around the world. It is your projects that inspired this book.

Rob McDonald dedicates this book to his children and hopes that the book can contribute, in some small way, to making the urban world they will inherit a little bit more verdant and soulful.

Contents

List of Figures

List of Tables

CHAPTER 1

The Urban Century

Abstract This century could rightly be called the urban century, with 2.4 billion more people expected in cities by 2050. This chapter presents the major choice facing planners: whether we want nature to play a central role in our urban future. We briefly present the central argument of this book that, paradoxically, cities need nature more than ever. Cities are the perfect forum for interaction, and it is this interaction that makes cities successful. Yet living in cities also carries with it an urban psychological penalty, as interacting with hundreds or thousands of people a day creates stress and mental illness. We believe that nature in cities can allow humanity to have all the benefits of a connected, urban life while alleviating the urban psychological penalty.

Keywords Climate change • Mental health • Natural features • New Urban Agenda • Urbanization • Urban growth

Hunter's Point

Hunter's Point existence is an accident of geography. Perched near the confluence of Newtown Creek and the East River, the peninsula became a dumping ground for waste from digging tunnels into Manhattan. The infill then became a bustling industrial site, with factories served by trucks and cargo ships. However, by the 1990s the economy of New York City had changed, and the site was economically abandoned and isolated.

R. McDonald, T. Beatley, *Biophilic Cities for an Urban Century*,
https://doi.org/10.1007/978-3-030-51665-9_1

Governor Mario Cuomo designated the land a future public park, but left the area fenced off, a vacant industrial wasteland. In 2012, Mayor Michael Bloomberg returned the focus to the site, and New York City faced a choice of what Hunter's Point should become (Doumar 2018; Davidson 2018).

New York City desperately needs new housing, as its population swells from 8.2 million in 2010 to a projected 9.0 million people in 2040 (The City of New York 2013). But the city is also trying to remake its shoreline, increasing the parkland for residents and reconnecting the city visually with the East River. After the massive flooding during Hurricane Sandy in October 2012, having a use for the site which was resilient to flooding became an essential focus of planning (Doumar 2018). Climate change will only worsen the risks of coastal flooding, as sea levels rise and rainstorms potentially get more intense.

In the end, New York City chose to achieve a little bit of both objectives: setting aside space for 5000 apartments while also adopting a plan to recreate a natural shoreline along Hunter's Point. Now, instead of concrete walls holding back the East River, marshes greet the daily tide waters. A little farther back from the water, a few trees and a grassy field offer a place to recreate and enjoy the stunning views of Manhattan's skyline. Curving concrete walkways allow for walks along the point, as well as future access by neighborhood residents, whose new building is slated for the most uphill part of the site (Davidson 2018). The redevelopment of Hunter's Point is still a work in progress, but the project is a good example that an expanded New York City, with close to a million new residents, need not be a gray, lifeless place. Instead, another city is possible, one in which nature is increasingly integrated into the fabric of a natureful, vibrant urban life (Fig. 1.1).

The choice that New York City faced—of how to develop while maintaining space for nature—is one that tens of thousands of cities and towns will face. This century will be the urban century, and we will witness the greatest migration in human history. By 2050, there will be 2.4 billion more people in cities. Humanity is doing the equivalent of building an urban area the size of New York City every nine weeks, over and over again (UNPD 2018). Will these cities be dull, lifeless places, or will they be lively, green ones? What use is nature, anyway, in the urban century?

Fig. 1.1 Hunter's Point South Park, New York, NY. (Photo credit: Hahn Darlin (CC BY 2.0))

Life in the Urban Century

Urban growth is a global mega-trend, one that will reshape many facets of human society. The trend will be most acutely felt in places like China (240 million new urbanites by 2050), India (180 million), and Nigeria (70 million) (UNPD 2018). But even in developed economies like the Unites States and the European Union, continued migration into large urban agglomerations will drive urban area expansion. While some declining industrial cities are losing population, the trend in most cities will be toward growth. *Homo sapiens* is already a majority urban species, and by 2050, it is projected that two out of every three humans will live in cities (UNPD 2018). Urban growth is the reality to which the fields of landscape architecture and sustainability science must respond if they are to remain relevant.

Some of the eye-popping increases in urban population are occurring simply because overall human population is increasing from 7.4 billion in 2015 to a forecasted 9.8 billion in 2050 (UNPD 2018). But urban population growth is also occurring because the fraction of people in cities is rising. The complex factors responsible for increasing the fraction of people in cities are detailed in Chap. 2. For now, it is enough to say that urban population growth seems a virtual certainty.

All these new urban dwellers will need some place to live. The most commonly cited forecast suggests that there may be 1.2 million km^2 of new urban area by 2030 (Seto et al. 2012). Humanity will, in other words, develop an area that, if you push it all together, could cover a country the size of South Africa. There is considerable uncertainty in forecasting future urban area (McDonald et al. 2018), and newer studies suggest a lower total area may be urbanized (Zhou et al. 2019). Perhaps most importantly, nothing in these forecasts is set in stone. Human choices about urban form and density will determine how much of our planet will be developed for urban land uses (Fig. 1.2).

Urban expansion will be greatest at the edges of existing metropolitan areas, on sites with flat land and close to existing transportation infrastructure like roads, rail lines, and ports. Historically, many cities have been located along coastlines and rivers, and most new urban growth will also occur in these locations. This matters for biodiversity because such natural habitat in these locations has high levels of productivity and biodiversity. Designing smartly the new urban area of the twenty-first century thus matters not just to human well-being but also to the preservation of biodiversity.

It is easy when discussing the urban century to imagine examples like New York City, large urban agglomerations housing tens of millions of people. But historical data and future forecasts suggest that urban population will be divided among many cities and towns of various sizes. Currently, almost half (49%) of urban dwellers are in cities of less than 500,000 people, a fraction that will remain roughly the same over the next few decades (UNPD 2018). Thus, landscape architecture and sustainability science need to define the role of nature in the urban century, not just for a handful of megacities but for thousands of smaller settlements. Smaller settlements face unique challenges, such as fewer economic resources and technical capacity, but also some advantages, including the ability to plan at smaller scales and lower population densities.

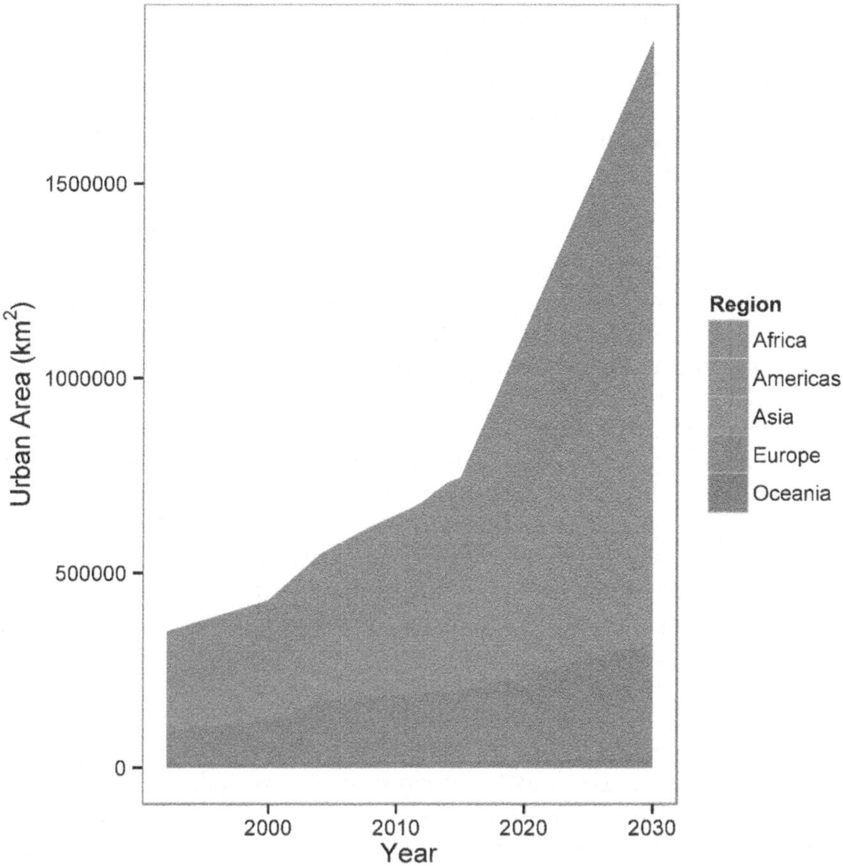

Fig. 1.2 Urban area growth over time for the major regions of the world. For details of the historical reconstruction and projection, please see the Nature in the Urban Century Assessment

Just how much area we urbanize also depends on the population density of new urban area. There is substantial regional variation, with cities in Asia having generally higher urban densities than in other places like North America and Australia. Correlated with this, cities with higher per-capita income usually have urban areas with lower population density (i.e., more built-up area consumed per capita). Over time, as economies get wealthier, urban agglomerations generally decrease in population density.

A lot of this trend toward decreased density has to do with the increased prevalence of car ownership in richer economies, which enables longer commutes and thus less dense patterns of settlement. But population density is also a function of transportation, zoning, and other planning decisions (Angel et al. 2012).

Designing the Cities of the Future

We are designing the cities of the future, today. How those cities are shaped is a choice, both of planners and, to perhaps a lesser degree, the general public. We must choose whether we want nature to play a role in our urban future. We can have an urban world that is gray and lifeless, or we can choose an urban world that is more verdant and alive.

In Quito in 2016, representatives of the world's governments at the UN Habitat III Conference agreed to the New Urban Agenda (UN Habitat 2016). It is the closest thing we have to a global, consensus statement on how the cities of the future should function. The document acknowledges the need for provision of adequate housing for all, and especially the need to improve the housing of the close to one billion people in informal settlements. But recognizing the broad links between urbanization and other aspects of human society, the New Urban Agenda also advocates for equity, sustainable economic development, and gender equality.

On environmental issues, the New Urban Agenda focuses on commitments to minimize the impacts of cities on the natural world while also ensuring a safe, clean environment for those in cities. The Agenda acknowledges the role of properly designed cities in ensuring urban stability and resilience. However, there is only a brief mention of natural ecosystems as a potential tool to reach resilience, and a commitment to providing equitable access to parks.

In this book, we argue that nature can play and must play a much bigger role in achieving the ambitious goals of the New Urban Agenda. We believe that natural features can play a broader role than mere aesthetics or risk reduction, that they are essential to the success of the urban century.

We acknowledge, of course, that there are many other important agendas that must be considered as humanity designs its urban home. For instance, those advocating for compact, walkable cities will stress the importance of density and how it can lead to less transportation energy use and hence lower greenhouse gas emissions, as well as improved physical health. Smart city advocates will argue for the role of new technology in increasing efficiency and productivity of urban economies. And looming over all urban planning decisions now, including that at Hunter's Point, is the potential impact of climate change. Cities must prepare for rising seas, worsening storms, insecure water supplies, hotter heat waves, and more intense wildfires, among other climate change risks. We do not mean for the agenda of nature in cities, the central focus of this book, to displace these other agendas, but rather to be complementary, enabling their success.

This Book

This book argues that, paradoxically, at their moment of triumph and fastest growth, cities need nature more than ever. Cities have been called our greatest invention, a perfect forum for interaction. As shown in Chap. 2, it is this extreme potential for interaction that makes cities successful. Productivity, innovation, and creativity are at their peak in cities, the denser the better. Yet cities are also unnatural. As discussed in Chap. 3, life in a dense city also carries with it an urban psychological penalty, as interacting with hundreds or thousands of people a day creates stress and mental illness (Fig. 1.3).

We believe that nature in cities can help humanity have its cake and eat it too, allowing humanity to have all the benefits of a connected, urban life while alleviating the urban psychological penalty. As discussed in Chap. 4, studies have shown that even a brief visit to a park or other open space can decrease stress, increase attention and cognitive function, and even reduce the symptoms of children with attention deficit disorder. Even a view of nature from a window has been shown to deliver benefits, from increased domestic harmony to faster patient recovery in hospitals.

We then turn to explaining *how* cities can design and plan for including nature in their territory. Our focus is on the philosophy and practice of biophilic design, the incorporation of nature into the built environment. While there are other planning processes and design philosophies that similarly aim to create nature-based solutions for cities, we believe that biophilic design, as one of the most mature and expansive philosophies, provides a good guide for how cities can have all the benefits of nature while alleviating the urban psychological penalty. We begin by presenting best practices from biophilic design (Chap. 5) that allow neighborhoods to incorporate nature, in myriad ways, into their urban fabric. These

Fig. 1.3 Many cities contain gray, lifeless landscapes. One famous example is the Los Angeles River (top), which for much of its length is channelized. (Photo credit: David Wilson (CC BY 2.0). Other cities have chosen to reincorporate nature into their urban landscape. Seoul, for instance, revitalized the once channelized Cheonggyecheon River into a green urban linear park (bottom). Photo credit: Brian Kusler (CC BY 2.0))

principles are then illustrated with case studies of several exciting new biophilic design projects (Chap. 6). We close the book by arguing in Chap. 7 that for our urban world to thrive, we must radially reimagine what nature can mean for the city.

References

Angel, Shlomo, Alejandro M. Blei, Daniel L. Civco, and Jason Parent. 2012. *Atlas of urban expansion*. Cambridge, MA: Lincoln Institute of Land Policy.

Davidson, J. 2018. New York City's newest park redeems a wasteland. *New York Intelligencer*, August 30.

Doumar, K. 2018. A storm-resilient park in Queens. *CityLab*.

McDonald, R., M'Lisa Colbert, M. Hamann, R. Simkin, B. Walsh, F. Ascensao, M. Barton, K. Crossman, M. Edgecomb, T. Elmqvist, A. Gonzalez, B. Guneralp, D. Haase, O. Hillel, Kangning Huang, D. Maddox, A.V. Mansur, H. Pereira, J.R. Pierce, R. Weller, K. Seto, M. Tan, and C. Ziter. 2018. *Nature in the urban century: A global assessment of where and how to conserve nature for biodiversity and human wellbeing*. Washington, DC: The Nature Conservancy.

Seto, K., B. Guneralp, and L. Hutyra. 2012. Global forecasts of urban expansion to 2030 and direct impacts on biodiversity and carbon pools. *Proceedings of the National Academy of Science* 109 (40): 16083–16088.

The City of New York. 2013. *New York city population projections by age/sex & borough, 2010–2040*. New York, NY: Department of City Planning, The City of New York. https://www1.nyc.gov/assets/planning/download/pdf/data-maps/nyc-population/projections_report_2010_2040.pdf.

UN Habitat. 2016. *Habitat III: New urban agenda*. Quito: UN Habitat. http://citiscope.org/sites/default/files/h3/Habitat_III_New_Urban_Agenda_10_September_2016.pdf.

UNPD. 2018. *World urbanization prospects: The 2018 revision*. New York, NY: United Nations Population Division.

Zhou, Yuerong, Alvin C.G. Varquez, and Manabu Kanda. 2019. High-resolution global urban growth projection based on multiple applications of the SLEUtH urban growth model. *Scientific Data* 6 (1): 34.

CHAPTER 2

Cities as Quintessentially Human

Abstract Cities are our greatest invention, an arrangement of homes and businesses that drastically increases the speed of interaction. This interaction then leads to numerous benefits, which is the primary subject of this chapter. We discuss first agglomeration economics—the benefits to firms and people productivity of being in clusters (agglomerations). Our discussion follows the traditional classification of benefits into sharing, matching, and learning. Then, we discuss research that shows that proximity also increases the possibilities for consumption, with a brief case study of restaurant variety in rural and urban areas. Finally, the chapter ends by arguing that the demographic forecasts are clear, urbanization is coming, and humanity therefore must turn its focus to the crucial question of what kind of urban world we want to live in.

Keywords Agglomeration • Consumption benefits • Learning • Matching • Migration • Production benefits • Sharing • Urbanization

Our Greatest Invention

The authors of this book live in urban areas, and love the variety of urban life, the "moveable feast" (Hemingway 2014) of sights and sounds and experiences. One of us lives in the town of Charlottesville, VA, a vibrant college town which fills every fall with students from around the world. They have all gathered in one place for learning, minds coming together

R. McDonald, T. Beatley, *Biophilic Cities for an Urban Century*,
https://doi.org/10.1007/978-3-030-51665-9_2

to exchange ideas and insights. One of us lives in the city of Washington, DC, the sprawling national capital of the United States. While five million people live in its metro area, much of the political action happens in a few square miles in the city's core. Here, policymakers and lobbyists congregate to make laws and policies, while bureaucrats huddle together to figure out how to implement them. It is this human interaction that makes both of our urban lives so varied and pleasing.

Aristotle famously said that "man is by nature a social animal" (Aristotle circa 330 BC). He meant by this that seeking interaction with other humans in a shared society is an essential part of our human nature. In cities, we are creating the perfect space for interaction, a form of living that facilitates and accelerates it. Seen from this perspective, creating our cities is fulfilling a deep-seated need for interaction. Cities are quintessentially human, a reflection of our essential nature.

The economist and urbanist Ed Glaeser once called cities "our greatest invention" (Glaeser 2012). The core innovation of urban areas, refined over the centuries, is the arrangement of our homes and businesses to increase the speed of interaction. This interaction then leads to numerous benefits, which is the primary subject of this chapter. While our discussion in this chapter is framed around the economic benefits of increased interaction in urban areas, it is important to recognize that interaction in cities can be viewed from many other perspectives. When Jane Jacobs (1961) wrote lovingly about the "ballet of the good city sidewalk" near her home on Hudson Street in Greenwich Village in Manhattan, she is marveling at the charm and coordination of all these urban interactions. William Whyte (1980) famously studied the way the physical environment, a city's plazas and small parks, shape how social interactions occurs. More broadly, increased personal freedom often goes along with the potential for more interaction. In medieval Germanic law, there was a principle that *Stadtluft macht frei* ("urban air makes you free"), and residence in a city was enough to free serfs from the dominion of their feudal lord. This same phrase has been employed by modern writers, talking about how urban life offers greater freedom than rural areas from the constraints of custom (Glaeser 2016). All these various perspectives on the moveable feast that is urban life follow from the increased interaction in cities.

In our lives in Charlottesville and in Washington, it is this potential for interaction that makes our cities great. Without the bustle of new ideas being exchanged, universities would cease to function, and politicians would cease writing laws. This chapter is about humanity's greatest invention, our cities. We discuss why people are moving to cities, and the many benefits of proximity. We then consider whether there is any alternative to urban life, or if the phenomenal growth predicted for urban areas is in some sense inevitable.

Why Is Everyone Moving to Cities?

We have made a big deal in this book about the urban century, the rapid growth of urban areas in the second half of the twentieth century and predicted to continue to 2050. But *why* is everyone moving to urban areas, anyway? This is a complex question, with no single answer. There are sometimes factors that push rural dwellers to leave their homes and migrate to cities. Sometimes this *push factor* is the collapse of a rural economy, as when a disease wiped out potato crops across Ireland, leading to widespread famine and poverty. Many of these farmers abandoned their fields and moved to cities, either nearby (Dublin) or internationally (Walsh 1999).

In this book, we follow many historians of urban growth by putting greater emphasis of factors that pull rural dwellers into cities. A *pull factor* can be anything that makes life in cities more appealing than in rural areas (Montgomery et al. 2003). Pull factors are necessary to explain urbanization, for without pull factors the collapse of a rural economy, for instance, might lead mostly to rural-to-rural migration. Many pull factors are a direct consequence of the increased potential for interaction that urban areas afford (Bettencourt et al. 2007; Bettencourt 2013; Bettencourt et al. 2010). Whole books have been written on the positive benefits of proximity in cities, so our review will be necessarily brief, citing more detailed texts that readers can refer to if interested.

Urban Benefits to Production

One common way economists study urbanization is by cataloging and quantifying how proximity benefits individuals and firms. This is sometimes called agglomeration economics—the benefits to firms and people of being in clusters (agglomerations). There is a voluminous literature on this topic, with influential reviews by Fujita and Thisse (2002) and Puga (2010). In the literature, there is agreement that there are multiple pathways by which agglomeration benefits flow. One useful classification is dividing benefits into sharing, matching, and learning (Duranton and Puga 2004; Andersson et al. 2007).

Sharing of infrastructure resources is possible when people and firms are in proximity. It is more cost-effective to build transportation infrastructure, like roads or railroads, when multiple people can use it. For longer-distance trade, facilities like ports or airports are essential, but their high cost of construction means it would be difficult for any single user to finance construction (Duranton 2008). Infrastructure can also be shared within a firm. If goods can be produced more cost-effectively by sharing machinery among different production processes, a firm may find that concentration of production is preferable to dispersed operations (Andersson et al. 2007).

One example of sharing might be in Singapore (Sheffi 2013). The city's harbor was a naturally sheltered port, in a location of strategic importance for trading, and it was this that led to the city's initial growth. However, major government investment has been required to continually upgrade the port so it could accommodate the ever-growing cargo ships. No one firm could afford to invest in such a resource on its own, but literally thousands of firms use the port. Singapore is one of the world's best examples of a logistics hub, a location where sharing benefits are key to economic productivity (Fig. 2.1).

The proximity of firms and people also enable *matching* (Andersson et al. 2007; Venables 2010; Wheeler 2001). Matching occurs when two actors find a mutually agreeable collaboration. It takes time and money for firms to find qualified workers. This cost is reduced when multiple firms in the same sector are co-located, which leads to a large pool of workers qualified to work in that sector. Such a large pool of workers makes it easier for firms to find qualified workers. Firms also benefit when they can quickly find other firms to match with. For instance, when firms that make electronic components for phones are co-located with factories that assemble phones, proximity aides in negotiating a contract and collaboration, which reduces transaction costs.

A great example of matching occurs in Hollywood, center of American movie and television production. Making such video requires specialized skills such as writing, acting, and directing. It is easier for firms to find good employees with needed skills if they cluster near other similar firms. It is also easier for employees to find jobs, moving between firms as gigs end (Fig. 2.2). Empirical evidence suggests that larger pools of workers

Fig. 2.1 A port in Singapore, a resource shared among many firms. (Photo credit: Timothy Beatley)

facilitate better matching, and having more skilled employees leads to greater productivity (Harmon 2013).

Finally, the proximity of people facilitates *learning* (Wheeler 2006; Lee and Rodríguez-Pose 2014; Glaeser and Resseger 2010). Research shows that many kinds of knowledge are best transmitted through direct interaction. This benefit of proximity still persists despite the rise of technologies that allow communication at a distance, such as the telephone and the Internet (Von Hippel 1994). Larger cities generally have higher rates of innovation, as measured by metrics such as patent generation (Bettencourt et al. 2007), supporting the idea that increased interaction leads to more learning. One individual's information spills over to others, creating what economists call a positive externality. To the students and academics gathering in Charlottesville, proximity enables increased information exchange that benefits everyone.

The most studied example of learning is Silicon Valley (Kenney 2000). The rise of the computing industry in Silicon Valley has been traced by historians to, in part, the high rate of personal interaction among workers,

Fig. 2.2 Hollywood, a place where matching between employers and employees is key. (Photo credit: BenSherman (public domain))

which has led to an accelerated rate of knowledge generation (Fallick et al. 2006). The multiple technology companies have workers who connect socially, sharing ideas and building off each other's innovations. The Silicon Valley culture of rapid job-hopping also increases the information sharing among firms. Interestingly, many of the large firms in Silicon Valley attempt to limit such information transfer by forcing workers to sign non-compete agreements that limit such job-hopping. Such non-compete agreements can be seen as an attempt to prevent the positive externality of information sharing and keep the worker's information as the private property of the firm (Fig. 2.3).

These three pathways add up to substantial agglomeration benefits. Urban wages increase with city size, as do company profits. Larger cities have more innovation as well. Cities, and the increased opportunities for interaction therein, are indeed significant boosts to overall economic productivity.

Fig. 2.3 Downtown San Jose, one end of the Silicon Valley corridor. Proximity of technology firms in Silicon Valley allows for learning of workers across companies. (Photo credit: Darshan Karia (public domain))

Urban Benefits to Consumption

In the past, urban economics research focused on the benefits of proximity to production. In recent decades, however, research has shown that equally important is the way proximity increases consumption (Glaeser et al. 2001; Shapiro 2006). A city has many amenities and services that are not as frequently available in rural areas. Stores, museums, and theaters are all examples of amenities that are available in greater variety in cities. Just as proximity has benefits for production, so it has benefits for consumption.

One quantitative example of urban agglomeration benefits to consumption is the number of restaurants (Couture 2013). A rural location in the United States might have less than 10 restaurants within a 45-minute drive. A resident of Des Moines, Iowa, has 800 restaurants to choose from within the same driving distance. Residents of a megacity like Los Angeles and New York typically have more than 20,000 restaurants within a 45-minute drive to choose from. The catch is that this urban variety comes at a price. The greater rent in urban areas leads to generally more expensive restaurants.

One bit of evidence for urban benefits to consumption comes from the phenomenon of reverse commuting. Rents in cities have grown faster than wages, suggesting that workers' interest in living in cities has risen faster than firms' interest in locating in urban areas (Glaeser et al. 2001). If firms are outside the city center but workers are not, that leads to a "reverse commute." In Philadelphia, for instance, two in five workers reverse commute, while in places like Detroit, more than 50% of workers reverse commute (Eichel and Budick 2018). Note that there can be multiple causes of reverse commuting, beyond the choice of consumers to locate in cities for the increased consumption possibilities. For instance, a spatial mismatch of housing and jobs can force people to reverse commute.

An Urban World, Ready or Not

As discussed in Chap. 1, the last few decades' urban growth is forecast to continue. Close to three billion new urban residents are expected by 2050 (UNPD 2018). While the exact number of additional urban residents forecast varies by study, demographers are almost unanimous in predicting further urban growth. Why are demographers so certain? After all, life in cities involves trade-offs, including higher rent and cost of living, as well as increased congestion.

Demographers are so confident in predicting urban growth because in the past urbanization has almost invariably been associated with economic development (Duranton 2008). The process of urbanization is often measured by the percentage of the total population living in cities (UNPD 2018). Mali, a low-income country according to UN classifications, is only 44% urbanized. The average urbanization for all such least developed countries is 35%. Conversely, Germany, a high-income country, is 78% urbanized. The average urbanization for all high-income countries is 82%. The correlation between urbanization and economic development also holds over time. Real gross domestic product per capita in the United States rose from around $5000 in 1900 to $52,000 in 2016 (Maddison 2010; BEA 2019). Over the same time period, urbanization rose from 40% to 83% (Boustan et al. 2013; UNPD 2018). Given the amount of economic development forecast for developing countries, continued urbanization in these countries seems almost assured.

It is also not clear what the alternative would be to urbanization. Some countries have feared the changes urbanization might bring and tried to limit it. These policies generally have failed to ultimately limit urbanization, although they may have slightly delayed it (cf., Jones and Corbridge 2010). An example might be the Chinese system of ***hukou***, or family registration (Afridi et al. 2015). In China, a citizen is designated as either rural or urban based upon their place of birth. In practice, it has been difficult for those with rural designation to get the full legal permits to live and work in urban areas. A migrant worker in a city like Shanghai may therefore not be entitled to full access to public services. The ***hukou*** system appears to have only modestly limited rural to urban migration. Rural migrants have still moved to cities in search of jobs, creating a large class of migrant workers in cities without full public services. Recognizing some of these issues, the Chinese government has begun reforming and loosening the ***hukou*** system to make it easier to switch designations (Chan 2019).

Another hope of those who fear an urban planet is that the widespread availability of the Internet and virtual communications could slow the rise in urban populations. The argument is that if cities are all about interaction, and interaction at a distance gets cheaper and easier, then people will have less of an incentive to settle in urban areas. Similar arguments have been made about driverless cars and other such technologies that reduce the costs of transportation.

However, the empirical evidence suggests the rise of the Internet and other means of communication at a distance have been associated with increased urbanization. The Internet appears to be a complement for face-to-face interactions, rather than a substitute (Gaspar and Glaeser 1998). Workers may more frequently work at home, but still need to periodically meet in one location for more in-depth communication. In fact, urban residents make more use of the Internet than rural ones, further increasing their advantage over rural dwellers in the pace of interactions.

Closing

Urbanization is coming, whether as a global society we are ready for it or not. It is wiser for policymakers to plan for the coming urban century than it is to fight to delay its arrival. The key question then becomes, what do we want those cities to look like? This is the key question humanity must answer as we move into the urban century.

We have tried to show in this chapter how interaction is central to cities. It is interaction that is behind the multitude of urban agglomeration benefits. One might be tempted even to ask how cities should structure themselves to maximize interaction. However, as we show in the next chapter, interaction and proximity have a dark side. The very thing that makes cities unique also contains a deep challenge to humanity.

References

Afridi, Farzana, Sherry Xin Li, and Yufei Ren. 2015. Social identity and inequality: The impact of China's hukou system. *Journal of Public Economics* 123: 17–29.

Andersson, Fredrik, Simon Burgess, and Julia I. Lane. 2007. Cities, matching and the productivity gains of agglomeration. *Journal of Urban Economics* 61 (1): 112–128.

Aristotle. circa 330 BC. *Politics*.

BEA, US. 2019. NIPA Table 7.1. US Bureau of economic analysis.

Bettencourt, Luís M.A. 2013. The origins of scaling in cities. *Science* 340 (6139): 1438–1441.

Bettencourt, L., J. Lobo, D. Helbing, C. Kuhnert, and G. West. 2007. Growth, innovation, scaling, and the pace of life in cities. *Proceedings of the National Academy of Science* 104 (17): 7301–7306.

Bettencourt, Luís M.A., José Lobo, Deborah Strumsky, and Geoffrey B. West. 2010. Urban scaling and its deviations: Revealing the structure of wealth, innovation and crime across cities. *PloS One* 5 (11): e13541.

Boustan, Leah Platt, Devin Bunten, and Owen Hearey. 2013. *Urbanization in the United States, 1800–2000.* National Bureau of Economic Research.

Chan, Kam Wing. 2019. China's hukou system at 60: Continuity and reform. In *Handbook on Urban Development in China*, 59–79. Edward Elgar.

Couture, Victor. 2013. *Valuing the consumption benefits of urban density.* Berkeley: Haas School of Business, University of California. https://pdfs.semanticscholar.org/8b2f/12e9848b149f9e31eccf971f321b2b89b904.pdf.

Duranton, Gilles. 2008. From cities to productivity and growth in developing countries. *Canadian Journal of Economics/Revue canadienne d'économique* 41 (3): 689–736.

Duranton, Gilles, and Diego Puga. 2004. Micro-foundations of urban agglomeration economies. *Handbook of Regional and Urban Economics* 4: 2063–2117.

Eichel, L., and S. Budick. 2018. *Reverse commuting in Philadelphia mirrors other large cities.* Washington, DC: Pew Trusts. https://www.pewtrusts.org/en/research-and-analysis/articles/2018/10/09/reverse-commuting-in-philadelphia-mirrors-other-large-cities.

Fallick, Bruce, Charles A. Fleischman, and James B. Rebitzer. 2006. Job-hopping in silicon valley: Some evidence concerning the microfoundations of a high-technology cluster. *The Review of Economics and Statistics* 88 (3): 472–481.
Fujita, M., and J.-F. Thisse. 2002. *Economics of agglomeration: Cities, industrial location, and regional growth.* Cambridge: Cambridge University Press.
Gaspar, Jess, and Edward L. Glaeser. 1998. Information technology and the future of cities. *Journal of Urban Economics* 43 (1): 136–156.
Glaeser, E. 2012. *Triumph of the city: How our greatest invention makes us richer, smarter, greener, healthier, and happier.* New York, NY: Penguin Books.
———. 2016. City air makes you free. *New Boston Post.*
Glaeser, Edward L., and Matthew G. Resseger. 2010. The complementarity between cities and skills. *Journal of Regional Science* 50 (1): 221–244.
Glaeser, Edward L., Jed Kolko, and Albert Saiz. 2001. Consumer city. *Journal of Economic Geography* 1 (1): 27–50.
Harmon, Nikolaj A. 2013. Are workers better matched in large labor markets? *Job market paper, Princeton University.*
Hemingway, Ernest. 2014. *Moveable feast: The restored edition.* New York, NY: Simon and Schuster.
Jacobs, J. 1961. *The death and life of great American cities: The failure of town planning.* New York: Random House.
Jones, Gareth A., and Stuart Corbridge. 2010. The continuing debate about urban bias: The thesis, its critics, its influence and its implications for poverty-reduction strategies. *Progress in Development Studies* 10 (1): 1–18.
Kenney, Martin. 2000. *Understanding silicon valley: The anatomy of an entrepreneurial region.* Palo Alto, CA: Stanford University Press.
Lee, Neil, and Andrés Rodríguez-Pose. 2014. Creativity, cities, and innovation. *Environment and Planning A* 46 (5): 1139–1159.
Maddison, Angus. 2010. Statistics on world population, GDP and per capita GDP, 1–2008 AD. *Historical Statistics* 3: 1–36.
Montgomery, M., R. Stren, B. Cohen, and H.E. Reed. 2003. *Cities transformed: Demographic change and its implications in the developing world.* Washington, DC: National Academies Press.
Puga, Diego. 2010. The magnitude and causes of agglomeration economies. *Journal of Regional Science* 50 (1): 203–219.
Shapiro, Jesse M. 2006. Smart cities: Quality of life, productivity, and the growth effects of human capital. *The Review of Economics and Statistics* 88 (2): 324–335.
Sheffi, Yossi. 2013. Logistics-intensive clusters: Global competitiveness and regional growth. In *Handbook of global logistics*, 463–500. Springer.
UNPD. 2018. *World urbanization prospects: The 2018 revision.* New York, NY: United Nations Population Division.
Venables, Anthony J. 2010. Productivity in cities: Self-selection and sorting. *Journal of Economic Geography* 11 (2): 241–251.

Von Hippel, Eric. 1994. 'Sticky information' and the locus of problem solving: Implications for innovation. *Management Science* 40 (4): 429–439.
Walsh, Brendan M. 1999. Urbanization and the regional distribution of population in post-famine Ireland. *Journal of European Economic History* 29: 109–127.
Wheeler, Christopher H. 2001. Search, sorting, and urban agglomeration. *Journal of Labor Economics* 19 (4): 879–899.
———. 2006. Cities and the growth of wages among young workers: Evidence from the NLSY. *Journal of Urban Economics* 60 (2): 162–184.
Whyte, W.H. 1980. *The social life of small urban spaces*. Washington, DC: The Conservation Foundation.

CHAPTER 3

Cities as Inhumane: The Urban Health Penalty

Abstract In this chapter, we explore ways in which urban environments and urban lifestyles are bad for human health. We discuss the urban health penalty historically, presenting a case study of how London solved its urban health penalty over the centuries. We then move on to discuss the aspects of the urban health penalty that remain in today's cities, with a focus on obesity and mental health disorders. Finally, we discuss in depth the potential causes of the urban psychological penalty. We argue that of these three environmental stressors are most amenable to mitigation by municipal officials, but that to some extent the urban psychological penalty is an inevitable consequence of the increased interaction inherent in urban life.

Keywords Density • Environmental stressors • Obesity • Schizophrenia • Stress • Urban health penalty

The commuters to New York City scurry like ants through Penn Station, row upon surging row of commuters all pushing forward in various directions. One can feel the hustle and hear the bustle that New York City is famous for, in a concentrated, distasteful dose. More than 600,000 people travel through this dingy underground train station each day (Kearney 2017). It is a place of stress, where people seem to go out of their way to disconnect from the chaos around them, hiding from the aural assault

R. McDonald, T. Beatley, *Biophilic Cities for an Urban Century*,
https://doi.org/10.1007/978-3-030-51665-9_3

behind ear buds playing a more soothing soundtrack, or reading on their smartphones while ignoring the sounds of the homeless begging for money or the drunks arguing with each other. While many people love New York City, nobody loves Penn Station. It perhaps epitomizes what people dislike about cities, how they can feel so full of life but so stressful.

Psychologists have found that there is a biological basis for all this stress. Every second in Penn Station, dozens or hundreds of commuters pass through your field of vision. Loud buzzers and angry shouts demand your attention, as your brain somehow tries to analyze what is a threat, what is potentially important information, and what can be safely ignored. To save valuable surface real estate, the current station was built underground with no natural light, only the harsh glare of artificial lighting. All these factions—the noise, the crowding, the unnatural lighting—are known by psychologists to contribute to stress and unease (Fig. 3.1).

Fig. 3.1 Pennsylvania station in New York City during a typical evening rush hour. More than 600,000 people every day move through the crowded, windowless station sunken underground. (Photo credit: frankieleon (CC BY 2.0))

In the previous chapter, we argued that cities are quintessentially human, that their form reflects a deep-seated desire for social interaction. But cities are also inhumane. Penn Station's hustle and bustle, and indeed urban environments in general, present people with far different environmental conditions than we evolved to handle (McDonald 2015). In this sense then cities are inhumane, for life in a city is sometimes not in accord with our organism's design and capacities.

In this chapter, we explore ways in which urban environments and urban lifestyles are bad for human health. We begin with a discussion of the urban health penalty historically. We then move on to discuss the aspects of the urban health penalty that remain today, with a focus on obesity and mental health disorders. Finally, we discuss in depth the potential causes of the urban psychological penalty, and how cities can help mitigate them.

The Urban Healthy Penalty Historically

Urban history can be seen, in part, as the struggle to make cities more humane. Health concerns in particular have driven many of the largest urban innovations over the last several centuries (McGranahan and Satterthwaite 2002). For much of urban history, urban dwellers had shorter lifespans than rural dwellers, a phenomenon called the urban health penalty. Scholars studying eighteenth- and nineteenth-century English demography coined the term "urban health penalty" to describe the reality that mortality rates, particularly from communicable diseases like typhoid, were substantially higher in urban than in rural areas (Kearns 1988). Urban infant mortality rates were also higher than rural mortality rates up until the late nineteenth century. Scholars studying United States cities in the same time period have found a similar urban health penalty (Vlahov and Galea 2002).

London may serve as the classic example of the urban health penalty (Landers and John 1993). Between 1500 and 1800 the population of London grew from around 50,000 to 960,000. However, up until the very end of this period the absolute number of deaths exceeded births. The mortality rate was so high that extra immigrants were needed every year just to make up for the scores of residents who perished from disease. For instance, in 1730, the crude death rate of 48.6 per thousand was more than 50% higher than the national average death rate of 31.4 per thousand, a dramatic empirical example of the urban health penalty.

In the first few decades of the 1800s, most deaths in London came from communicable diseases. One out of every three deaths were of children, most caused by diseases like whooping cough, measles, scarlet fever, and diphtheria. Tuberculosis, then called consumption, was the cause of one out of every four deaths, primarily striking adults. Another 6.9% of deaths were from "fevers," often caused by typhoid and typhus, while 5.4% of deaths came from smallpox. Cholera deaths could be minor in one year, and then reach epidemic levels in the next. As shocking as this high death rate is to us today, at the time it was seen as a slight improvement over the late 1700s. Indeed, by 1820 the urban health penalty had narrowed, with 26.7 deaths per thousand in London and 23.8 deaths per thousand in the rest of the country (Hardy et al. 1993).

This pattern of improving health outcomes was replicated in cities around the world. The urban health penalty, as measured by the difference in death rates between urban and rural areas, narrowed. One useful framework for understanding how cities have solved environmental challenges is called the urban environmental transitions (UET). As defined by McGranahan and colleagues (McGranahan 2007, McGranahan and Satterthwaite 2002), the UET argues that often cities have dealt with environmental challenges in a predictable order. First, cities focused on acute, neighborhood-scale challenges such as laying pipes to houses to provide clean drinking water and remote sewage. Second, cities moved on to less acute, more citywide issues such as air quality. Sometimes they merely displace the environmental burden elsewhere outside the city center, as when cities improve air quality merely by moving coal-fired power plants farther away. Finally, in just the last century, cities have moved on to trying to address their contribution to regional or global problems such as acid rain or climate change.

Dealing with sanitation and water quality is one of the first steps in the UET. London and other cities in England began to tackle this problem in the nineteenth century (Pike 2018). In 1855, the Metropolis Local Management Act required London's civil vestries—local neighborhood governance committees organized around parishes—to appoint medical officers to improve public health. In that same year, the Metropolitan Board of Works was formed. Joseph Bazalgette was appointed its first chief engineer, and one of his first major acts was drafting a plan for an integrated sewer system for London. However, little progress was made until after the Great Stink of June 1858. The Great Stink occurred when hot weather and the growing mass of untreated human waste in the River

Thames led to smells so bad that that it disrupted the work of Parliament, with lawmakers too sickened to work. A planned pleasure cruise by Queen Victoria had to be abandoned within minutes, because the smell was unbearable. Life in much of central London was profoundly disrupted (Ackroyd 2011). The smell became so bad that city and national leaders realized they had a significant problem and put substantial money toward completing the sewer system, which was finished by 1875 (Pike 2018).

The sanitation improvements, coupled with improvements in the nutrition and diet of the urban poor, led to an improvement in human health. Globally, drinking water and sanitation systems helped decrease mortality rates from water-borne communicable diseases such as typhoid and cholera. One scholar has called this new clean urban form the "Sanitary City" (Melosi 2008), and it was arguably the single most important step in alleviating the urban health penalty. However, this varied depending on the disease. For water-borne diseases like typhoid and cholera, the development of safe drinking water supplies and separate "sanitary" sewer systems were key. For tuberculosis, improvement in nutrition, medical care, and air quality all appear to have helped reduce the death rate from the disease. Other diseases like smallpox were conquered through vaccines.

We want to acknowledge, of course, much work still remains to be done to solve urban sanitation problems: 15% of urban dwellers globally still lack safely managed drinking water sources and 16% still lack access to at least basic sanitation (WHO and UNICEF 2017). In the twenty-first century, there are hundreds of millions of urban dwellers in neighborhoods still trying to solve the problem London tackled in the nineteenth century, many of them dwelling in informal settlements in developing countries.

Cities began later to address problems of ambient air pollution. In London, for instance, the infamous Great Smog of December 5–8, 1952, was severe enough that visibility was reduced so much that cars had to turn their headlights on during the daytime. The opera had to cancel shows since participants inside had trouble seeing the stage (Greater London Authority n.d.). By one estimate, 12,000 people died from the Great Smog (Bell et al. 2004). One cause of the Great Smog was domestic coal burning for heating and cooking, which the municipal government began quickly phasing out. The national Clean Air Act of 1956 began reducing industrial emissions, the other main cause of London smog. Now, more than 60 years after the Great Smog, London's concentrations of suspended particulate matter are one-fifth of what they were (Fouquet 2011).

Globally, these and similar air pollution emission controls have dramatically cleaned up the air. In the United States, particulate matter emissions (PM_{10}) fell by 57% between 1980 and 2016 (EPA 2016). Much work of course remains to be done, as in the case of water pollution. Ambient outdoor particulate pollution still kills globally more than three million people per year, in both rural and urban areas (Lim et al. 2012). Indeed, poor air quality events in cities like Beijing and Delhi are in some ways historical analogies of London's Great Smog, with similarly dire health outcomes (Fig. 3.2).

The good news is that while many developing country cities continue to struggle with dirty water and air, urban history does not have to be destiny. Recent papers have suggested that cities in the developing countries are not sequentially moving through the UET as did first world cities (Marcotullio 2007). Rather, Asian cities at least seem to be moving through environmental transitions faster, sometimes dealing with multiple transitions at once. These transitions are also happening at a lower level of average per-capita income. There is something hopeful in this finding, for it means that cities in the developing world can more quickly solve the underlying causes of much of the urban health penalty.

Fig. 3.2 Dense smog blocks out the sun in New Delhi, November 2019. Air pollution concentrations in cities like New Delhi now rival or exceed the London Great Smog of 1952. (Photo credit: Ninara (CC BY 2.0))

The Urban Health Penalty Today

The urban health penalty has now in aggregate disappeared, with mortality rates for urban dwellers now lower than those for rural dwellers (Leon 2008). For instance, in developing countries, the Demographic and Health Surveys (DHS) show an infant mortality rate of 86 deaths per 1000 live births in rural areas, 75 for the urban poor, and 56 in the urban non-poor (Vlahov et al. 2005). This means that even the urban poor, more than 800 million of whom live in informal or slum settlements (UN Habitat 2016), are generally healthier than rural dwellers. In the past, immigration into cities was needed to overcome high mortality rates and help cities grow. Currently, most urban growth globally comes from intrinsic increase, as births to urban residents exceed deaths.

If cities are our greatest invention, then version 2.0 (the Sanitary City) has been a vast improvement on version 1.0 in terms of health. However, two major facets of the urban health penalty remain. For both facets, disease rates are higher in urban dwellers than in rural dwellers.

First, on average, obesity and its associated diseases continue to be worse in urban than in rural areas (Allender et al. 2008). Worldwide, obesity now affects more than 600 million people, a doubling since 1980 (WHO 2017). Obesity occurs most frequently when there is an increased intake of calorie-rich foods and less physical activity (Swinburn et al. 2011). Both phenomena are more common in cities, where greater incomes allow more calorie intakes and a larger fraction of jobs are sedentary. Interestingly, the gap in obesity between rural and urban areas appears greatest in the poorest countries, presumably because in relative terms, the urban income advantage is greatest there. Conversely, in many developed countries, incomes in both rural and urban areas are sufficient to obtain enough calories of food, and there is no gap in obesity rates between rural and urban areas (Bixby et al. 2019). Indeed, in the United States, there is evidence that dense urban living is associated with more physical activity and lower body mass index (BMI) (Ewing et al. 2003).

However, it is unclear whether it is cities *per se* that are responsible for obesity. That is, cities are associated in the developing world with greater income, which itself may lead to obesity, but urban form itself may not be that important. As evidence for this view, recent research (Bixby et al. 2019) suggests that much of the growth of obesity globally in recent decades is due to increases in rural obesity rates. The increase in rural BMI between 1985 and 2017 was 2.09 kg/m^2 and 2.10 among men and women, respectively. In urban areas, the increase was 1.35 kg/m^2 and 1.59 for men and women, respectively. In other words, the urban health penalty is decreasing, but not because cities are getting healthier! Rather, obesity rates in rural areas are rising to catch up with obesity rates in cities.

Many governments, in both urban and rural areas, are now conducting programs to reduce obesity rates. In London, for instance, a study found that obesity costs the city £883.6 million a year, meaning around 0.4% of the economy of the city is lost to obesity-related costs each year (Greater London Authority 2011). To combat obesity, London has increased bike lanes and taken steps to make bike commuting easier. They are also trying to make more accessible public swimming pools and gyms. A report by the Organization for Economic Cooperation and Development (OECD) (Smith 2011), however, suggests that physician/dietician counseling is the most cost-effective strategy for combating childhood obesity. One challenge then for municipal governments battling obesity is that the most cost-effective strategies must be administered by the health sector, which may not be under municipal direction.

Second, some aspects of mental health are worse in urban areas than in rural areas (Gruebner et al. 2017). This urban psychological penalty has been shown to be the reason for stress. Urban life has been associated with greater stress than rural life, as well as with changes to brain function (Lederbogen et al. 2011). The stress most people feel in places like Penn Station is real and clinically significant. One study of more than four million adults in Sweden (2004) found a significant increase in rates of psychosis and depression in higher densities in cities than in lower density in rural areas. Similar studies of large cohorts in Denmark (Mortensen et al. 1999, Pedersen and Mortensen 2001) and the United Kingdom (Newbury et al. 2016; Paykel et al. 2000) have found negative impacts of urban life on mental health. This urban psychological penalty is not just a first world problem. In China, migrants from rural to urban areas have greater rates of schizophrenia (Chan et al. 2015). In Sao Paulo, Brazil, rates of self-reported mental distress are greater in high-density than in low-density neighborhoods (Andrade et al. 2012).

There is also evidence that the effect of urban life on mental health seems to vary by age. In children, some studies have found an association between urban life and greater frequency of attention deficit hyperactivity disorder (ADHD) (Pillay et al. 1999), although other studies did not (Cuffe et al. 2005). In adults, by contrast, urban living is associated with great incidence of schizophrenia (Freeman 1994; Lewis et al. 1992) and psychosis (Sundquist et al. 2004).

The relationship between urban life and the urban psychological penalty is complex, varying by disease and by the urban cultural context (Pedersen and Mortensen 2001). However, the greater population densities found in cities do, on average, appear to be associated with higher incidences of specific mental health diseases. Maps of global population density show that many people live at densities where this urban mental health penalty will occur (McDonald et al. 2018). In 2000, a total of 2.5 billion people (41% of all people) lived at densities greater than 800 people per square kilometer, the threshold found in the Swedish study to increase the risk of depression and psychosis (2004). By 2020, there were 3.7 billion people above this threshold, or close to one out of every two people. If the effect found in the Swedish study is applicable elsewhere, then close to half of humanity is living at urban densities that significantly increase the risk of mental health study (McDonald et al. 2018).

Many cities around the world now have programs focused on improving mental health. In London, for instance, Mayor Sadiq Kahn recently launched the Thrive LDN campaign (Greater London Authority 2014). This communications campaign reminds Londoners how prevalent mental health issues are, with one of every four people in the city impacted by a mental health disorder. The campaign also encourages people to overcome the stigma that prevents honest discussion of mental health disorders, and to seek treatment as appropriate. How effective this kind of communication campaign will be in improving treatment of mental health diseases remains to be seen.

Causes of the Urban Psychological Penalty

The strongest remaining aspect of the urban health penalty thus seems to be the worse mental health of urban dwellers. Why does this urban psychological penalty occur? Three categories of causes have been suggested (cf., Vlahov and Galea 2002): compositional differences in urban and rural populations, and a breakdown in social cohesion in urban areas and urban environmental stressors. In this section, we discuss each potential causal pathway in detail, and then describe what cities can do to mitigate this causal pathway.

Compositional Differences

Cities have a different mix of populations than do rural areas. This compositional difference makes analyses comparing the mental health of urban and rural dwellers challenging, since so much else is different between the populations. Those of lower economic status are sometimes concentrated in certain cities or urban neighborhoods. So sometimes are marginalized racial or ethnic groups. Since these populations are sometimes more prone to mental health problems, their concentration in cities makes urban areas statistically correlated with poor mental health.

Other compositional differences make comparison of mental health between rural and urban dwellers difficult. Urban population tends to be younger and less likely to be married than rural dweller. Even within cities, denser neighborhoods tend to be younger and more frequently married. Since the incidence of mental health diseases varies by age and marital status, these compositional differences lead to differences in average mental health between rural and urban dwellers. Similarly, rural populations tend to be poorer than urban population, and income is positively correlated with mental health. Moreover, for reasons of income and access, many rural dwellers are less likely to get adequate health care, and physical disease is correlated with mental health.

Compositional differences are clearly part of the reason for a correlation between urban/rural status and some mental health diseases. Indeed, statistical analyses in some geographic areas have found that, for some diseases, the urban psychological penalty may be fully explained by compositional differences (Robinson et al. 2017). Whatever their importance, there is not much that a city can, or should, do to stop the occurrence of such compositional differences. However, a thorough understanding of the different populations within a city and their mental health challenges can help health officials design effective mental health programs. Rather than one psychological health penalty, there are multiple ones, experienced in different ways by different populations.

Social Cohesion Breakdown

In certain cities or urban neighborhoods, urban life (sometimes called "urbanicity") may lead to a breakdown in social cohesion, the willingness of members of a society to cooperate with each other in order to survive and prosper (Stanley 2003). It is unclear in the scientific literature if this breakdown in social cohesion is due to urbanicity *per se*, or whether it simply is statistically more likely to occur in cities for compositional reasons (Wirth 1938; Saunders 2003). On the one hand, urban life can lend itself to greater anonymity and independence than rural life, which some argue leads to alienation. On the other hand, cities tend to have more single parents, as well as more families with no extended family living nearby. Cities are also commonly a place with more transient or temporary residents. These compositional factors are correlated with low scores on measures of social cohesion. Whatever the cause of low social cohesion in cities, since social cohesion is linked to mental health, processes in cities that reduce social cohesion are associated with poorer mental health.

There is little a city government can do to promote social cohesion *per se*. Many existing programs on this issue focus on integrating specific marginalized communities, such as refugees, into the social life of a city or region. Efforts to improve social cohesion also sometimes focus on reducing other underlying causes, such as chronic unemployment. Sometimes investments in creating new urban gathering spaces such as playgrounds, public squares, or parks are justified in terms of increasing community cohesion. However, there is relatively little scientific evidence that these changes in urban form strongly affect measures of social cohesion.

Environmental Stressors

There are numerous environmental stressors in cities that are associated with poor mental health (Evans 2003). For instance, the greater ambient noise in urban environments—such as the din of Penn Station at rush hour—is associated with higher stress levels and mental health issues (Moudon 2009; Hammersen et al. 2016). Ditto for the greater visual stimulation in cities, the way we are constantly passing many other potential friends or foes in a crowded place like Penn Station (Chepesiuk 2009). Being exposed to artificial light during nighttime hours seems particularly damaging to normal circadian rhythms and can disrupt sleep. Similarly, nighttime noise can disrupt sleep, and sleep deprivation is linked to mental disease. Moreover, excess noise in general appears to affect HPA (hypothalamic-pituitary-adrenal) axis functioning, which is involved with regulation, the stress response in humans. Excess noise results in feelings of annoyance and anger, which increases stress. For all these reasons, noise pollution is correlated with the incidence of mental diseases like depression (van den Bosch and Meyer-Lindenberg 2019).

Another potentially important environmental stressor is air pollution (van den Bosch and Meyer-Lindenberg 2019). A variety of air pollutants have been studied, but the strongest evidence of a link between pollutant concentration and mental health is for particulate matter (PM). The smallest particles, smaller than 2.5 micrometers ($PM_{2.5}$), are potentially the most damaging, as they can be inhaled and enter the bloodstream, where they can interact with the brain. Some studies have correlated PM concentrations with the incidence of suicide and depression. Clearer in the scientific literature is the causal link between PM and some physical health diseases, such as cardiovascular disease and asthma. The occurrence of these physical health diseases would, of course, tend to be positively correlated with mental health diseases.

In contrast to other causal explanations, cities can do a lot to reduce environmental stressors. Many cities are beginning to realize that reducing noise pollution does more than avoiding a public nuisance. Rather, reducing noise pollution can be part of a public health strategy for a city, as the quieter environment improves people's mental health. Similarly, reducing air pollution has numerous public health benefits, such as reducing asthma attacks. It appears that increased mental health is likely to be a benefit of improved air quality.

The Irony of the Urban Health Penalty

Life in cities is associated with poor mental health. This appears particularly true at high population densities, at least when the crowding is with non-family members (Epstein 1981; Evans 1979). However, some degree of crowding is part of what makes cities humanity's "greatest invention" (Glaeser 2012). Higher densities and a faster pace of life are part of what makes cities economically successful. It is ironic that the same phenomenon, cities' capacity to increase interaction, is both what makes cities great and what makes cities mentally inhumane.

It is possible that cities can prevent some of the mental health penalty by a targeted campaign to reduce environmental stressors. It is our belief, however, that a mere reduction in environmental stressors will not be enough to close the mental health penalty. Rather, we believe that some degree of mental stress is inevitable when living at high population densities. The urban psychological penalty may not be as easily engineered away with gray infrastructure as were other facets of the urban health penalty such as the increased incidence of water-borne diseases. Instead, we must rethink how we design our cities, so we get the benefits of humanity's "greatest invention" while also having a psychologically healthy place to live. And that is where we believe nature can play an essential role, as we will discuss in the next chapter.

References

Ackroyd, Peter. 2011. *London under*. London: Random House.

Allender, Steven, Charlie Foster, Lauren Hutchinson, and Carukshi Arambepola. 2008. Quantification of urbanization in relation to chronic diseases in developing countries: A systematic review. *Journal of Urban Health* 85 (6): 938–951.

Andrade, Laura Helena, Yuan-Pang Wang, Solange Andreoni, Camila Magalhães Silveira, Clovis Alexandrino-Silva, Erica Rosanna Siu, Raphael Nishimura, James C. Anthony, Wagner Farid Gattaz, and Ronald C. Kessler. 2012. Mental disorders in megacities: Findings from the Sao Paulo megacity mental health survey, Brazil. *PloS One* 7 (2): e31879.

Bell, Michelle L., Devra L. Davis, and Tony Fletcher. 2004. A retrospective assessment of mortality from the London smog episode of 1952: The role of influenza and pollution. *Environmental Health Perspectives* 112 (1): 6.

Bixby, Honor, James Bentham, Bin Zhou, Mariachiara Di Cesare, Christopher J. Paciorek, and NCD Risk Factor Collaboration. 2019. Rising rural body-mass index is the main driver of the global obesity epidemic. *Nature* 569: 260–264.

Chan, Kit Yee, Fei-fei Zhao, Shijiao Meng, Alessandro R. Demaio, Craig Reed, Evropi Theodoratou, Harry Campbell, Wei Wang, and Igor Rudan. 2015. Urbanization and the prevalence of schizophrenia in China between 1990 and 2010. *World Psychiatry* 14 (2): 251–252.

Chepesiuk, Ron. 2009. Missing the dark: Health effects of light pollution. *Environmental Health Perspectives* 117 (1): A20.

Cuffe, Steven P., Charity G. Moore, and Robert E. McKeown. 2005. Prevalence and correlates of ADHD symptoms in the national health interview survey. *Journal of Attention Disorders* 9 (2): 392–401.

EPA. 2016. *Air quality-national summary*. Washington, DC: Environmental Protection Agency. https://www.epa.gov/air-trends/air-quality-national-summary.

Epstein, Yakov M. 1981. Crowding stress and human behavior. *Journal of Social Issues* 37 (1): 126–144.

Evans, Gary W. 1979. Behavioral and physiological consequences of crowding in humans. *Journal of Applied Social Psychology* 9 (1): 27–46.

———. 2003. The built environment and mental health. *Journal of Urban Health* 80 (4): 536–555.

Ewing, R., T. Schmid, R. Killingsworth, A. Zlot, and S. Raudenbush. 2003. Relationship between urban sprawl and physical activity, obesity, and morbidity. *American Journal of Health Promotion* 18 (1): 47–57.

Fouquet, Roger. 2011. Long run trends in energy-related external costs. *Ecological Economics* 70 (12): 2380–2389.

Freeman, Hugh. 1994. Schizophrenia and city residence. *The British Journal of Psychiatry* 164: 39–50.

Glaeser, E. 2012. *Triumph of the city: How our greatest invention makes us richer, smarter, greener, healthier, and happier.* New York: Penguin Books.

Greater London Authority. 2011. *Childhood obesity in London.* London: Greater London Authority.

———. 2014. *London mental health: The invisible costs of mental ill health.* London: Greater London Authority.

———. n.d. *50 years on: The struggle for air quality in London since the great smog of December 1952.* London: Greater London Authority.

Gruebner, Oliver, Michael A. Rapp, Mazda Adli, Ulrike Kluge, Sandro Galea, and Andreas Heinz. 2017. Cities and mental health. *Deutsches Ärzteblatt International* 114 (8): 121.

Hammersen, Friederike, Hildegard Niemann, and Jens Hoebel. 2016. Environmental noise annoyance and mental health in adults: Findings from the cross-sectional German Health Update (GEDA) study 2012. *International Journal Of Environmental Research and Public Health* 13 (10): 954.

Hardy, Anne, Thomas Hardy, and Hart Hardy. 1993. *The epidemic streets: Infectious disease and the rise of preventive medicine, 1856–1900.* Oxford: Oxford University Press.

Kearney, L. 2017. Evening rush hour a 'test' for NYC's Penn Station commuters—Amtrak. *Reuters*, July 10.

Kearns, Gerry. 1988. The urban penalty and the population history of England. In *Society, health, and population during the demographic transition*, 213–236. Stockholm: Almqvist and Wiksell International.

Landers, John, and Landers John. 1993. *Death and the metropolis: Studies in the demographic history of London, 1670–1830.* Vol. 20. Cambridge University Press.

Lederbogen, F., P. Kirsch, L. Haddad, F. Streit, H. Tost, P. Schuch, S. Wust, J.C. Pruessner, M. Rietschel, M. Deuschle, and A. Meyer-Lindenberg. 2011. City living and urban upbringing affect neural social stress processing in humans. *Nature* 474 (7352): 498–501. https://doi.org/10.1038/nature10190.

Leon, David A. 2008. *Cities, urbanization and health.* Oxford University Press.

Lewis, Glyn, Anthony David, S. Andréassson, and Peter Allebeck. 1992. Schizophrenia and city life. *The Lancet* 340 (8812): 137–140.

Lim, Stephen S., Theo Vos, Abraham D. Flaxman, Goodarz Danaei, Kenji Shibuya, Heather Adair-Rohani, Mohammad A. AlMazroa, Amann Markus, H. Ross

Anderson, Kathryn G. Andrews, Martin Aryee, Charles Atkinson, Loraine J. Bacchus, Adil N. Bahalim, Kalpana Balakrishnan, John Balmes, Suzanne Barker-Collo, Amanda Baxter, Michelle L. Bell, Jed D. Blore, Fiona Blyth, Carissa Bonner, Guilherme Borges, Rupert Bourne, Michel Boussinesq, Michael Brauer, Peter Brooks, Nigel G. Bruce, Bert Brunekreef, Claire Bryan-Hancock, Chiara Bucello, Rachelle Buchbinder, Fiona Bull, Richard T. Burnett, Tim E. Byers, Bianca Calabria, Jonathan Carapetis, Emily Carnahan, Zoe Chafe, Fiona Charlson, Honglei Chen, Jian Shen Chen, Andrew Tai-Ann Cheng, Jennifer Christine Child, Cohen Aaron, K. Ellicott Colson, Benjamin C. Cowie, Sarah Darby, Susan Darling, Adrian Davis, Louisa Degenhardt, Frank Dentener, Don C. Des, Karen Devries Jarlais, Mukesh Dherani, Eric L. Ding, E. Ray Dorsey, Tim Driscoll, Karen Edmond, Suad Eltahir Ali, Rebecca E. Engell, Patricia J. Erwin, Saman Fahimi, Gail Falder, Farshad Farzadfar, Alize Ferrari, Mariel M. Finucane, Seth Flaxman, Francis Gerry R. Fowkes, Greg Freedman, Michael K. Freeman, Emmanuela Gakidou, Santu Ghosh, Edward Giovannucci, Gerhard Gmel, Kathryn Graham, Rebecca Grainger, Bridget Grant, David Gunnell, Hialy R. Gutierrez, Wayne Hall, Hans W. Hoek, Anthony Hogan, H. Dean Hosgood III, Damian Hoy, Howard Hu, Bryan J. Hubbell, Sally J. Hutchings, Sydney E. Ibeanusi, Gemma L. Jacklyn, Rashmi Jasrasaria, Jost B. Jonas, Haidong Kan, John A. Kanis, Nicholas Kassebaum, Norito Kawakami, Young-Ho Khang, Shahab Khatibzadeh, Jon-Paul Khoo, Cindy Kok, Francine Laden, Ratilal Lalloo, Qing Lan, Tim Lathlean, Janet L. Leasher, James Leigh, Yang Li, John Kent Lin, Steven E. Lipshultz, Stephanie London, Rafael Lozano, Yuan Lu, Joelle Mak, Reza Malekzadeh, Leslie Mallinger, Wagner Marcenes, Lyn March, Robin Marks, Randall Martin, Paul McGale, John McGrath, Sumi Mehta, Ziad A. Memish, George A. Mensah, Tony R. Merriman, Renata Micha, Catherine Michaud, Vinod Mishra, Khayriyyah Mohd Hanafiah, Ali A. Mokdad, Lidia Morawska, Dariush Mozaffarian, Tasha Murphy, Mohsen Naghavi, Bruce Neal, Paul K. Nelson, Joan Miquel Nolla, Rosana Norman, Casey Olives, Saad B. Omer, Jessica Orchard, Richard Osborne, Bart Ostro, Andrew Page, Kiran D. Pandey, Charles D.H. Parry, Erin Passmore, Jayadeep Patra, Neil Pearce, Pamela M. Pelizzari, Max Petzold, Michael R. Phillips, Dan Pope, C. Arden Pope III, John Powles, Mayuree Rao, Homie Razavi, Eva A. Rehfuess, Jürgen T. Rehm, Beate Ritz, Frederick P. Rivara, Thomas Roberts, Carolyn Robinson, Jose A. Rodriguez-Portales, Isabelle Romieu, Robin Room, Lisa C. Rosenfeld, Ananya Roy, Lesley Rushton, Joshua A. Salomon, Uchechukwu Sampson, Lidia Sanchez-Riera, Ella Sanman, Amir Sapkota, Soraya Seedat, Peilin Shi, Kevin Shield, Rupak Shivakoti, Gitanjali M. Singh, David A. Sleet, Emma Smith, Kirk R. Smith, Nicolas J.C. Stapelberg, Kyle Steenland, Heidi Stöckl, Lars Jacob Stovner, Kurt Straif, Lahn Straney, George D. Thurston, Jimmy H. Tran, Rita Van Dingenen, Aaron van Donkelaar, J. Lennert Veerman,

Lakshmi Vijayakumar, Robert Weintraub, Myrna M. Weissman, Richard A. White, Harvey Whiteford, Steven T. Wiersma, James D. Wilkinson, Hywel C. Williams, Warwick Williams, Nicholas Wilson, Anthony D. Woolf, Paul Yip, Jan M. Zielinski, Alan D. Lopez, Christopher J.L. Murray, and Majid Ezzati. 2012. A comparative risk assessment of burden of disease and injury attributable to 67 risk factors and risk factor clusters in 21 regions, 1990–2013: A systematic analysis for the Global Burden of Disease Study 2010. *The Lancet* 380(9859):2224–2260.https://doi.org/10.1016/S0140-6736(12)61766-8.

Marcotullio, P. 2007. Urban water-related environmental transitions in Southeast Asia. *Sustainability Science* 2: 27–54. https://doi.org/10.1007/s11625-006-0019-0.

McDonald, R.I. 2015. *Conservation for cities: How to plan & build natural infrastructure*. Washington, DC: Island Press.

McDonald, Robert Ian, Timothy Beatley, and Thomas Elmqvist. 2018. The green soul of the concrete jungle: The urban century, the urban psychological penalty, and the role of nature. *Sustainable Earth* 1 (1): 3.

McGranahan, G. 2007. Urban transitions and the spatial displacement of environmental burdens. In *Scaling urban environmental challenges: From local to global and back*, ed. P. Marcotullio and G. McGranahan, 18–44. London: Earthscan.

McGranahan, G., and D. Satterthwaite. 2002. The environmental dimensions of sustainable development for cities. *Geography* 87: 213–226.

Melosi, M.V. 2008. *The sanitary city*. Pittsburg, PA: University of Pittsburg Press.

Mortensen, Preben Bo, Carsten Bøcker Pedersen, Tine Westergaard, Jan Wohlfahrt, Henrik Ewald, Ole Mors, Per Kragh Andersen, and Mads Melbye. 1999. Effects of family history and place and season of birth on the risk of schizophrenia. *New England Journal of Medicine* 340 (8): 603–608.

Moudon, Anne Vernez. 2009. Real noise from the urban environment: How ambient community noise affects health and what can be done about it. *American Journal of Preventive Medicine* 37 (2): 167–171.

Newbury, Joanne, Louise Arseneault, Avshalom Caspi, Terrie E. Moffitt, Candice L. Odgers, and Helen L. Fisher. 2016. Why are children in urban neighborhoods at increased risk for psychotic symptoms? Findings from a UK longitudinal cohort study. *Schizophrenia Bulletin* 42 (6): 1372–1383.

Paykel, E.S., R. Abbott, R. Jenkins, T.S. Brugha, and H. Meltzer. 2000. Urban–rural mental health differences in Great Britain: Findings from the National Morbidity Survey. *Psychological Medicine* 30 (2): 269–280.

Pedersen, Carsten Bøcker, and Preben Bo Mortensen. 2001. Evidence of a dose-response relationship between urbanicity during upbringing and schizophrenia risk. *Archives of General Psychiatry* 58 (11): 1039–1046.

Pike, David L. 2018. *Subterranean cities: The world beneath Paris and London, 1800–1945*. Ithica: Cornell University Press.

Pillay, A.L., P. Naidoo, and M.R. Lockhat. 1999. Psychopathology in urban and rural/peri-urban children seeking mental health care. *South African Journal of Psychology* 29 (4): 178–183.

Robinson, Lara R., Joseph R. Holbrook, Rebecca H. Bitsko, Sophie A. Hartwig, Jennifer W. Kaminski, Reem M. Ghandour, Georgina Peacock, Akilah Heggs, and Coleen A. Boyle. 2017. Differences in health care, family, and community factors associated with mental, behavioral, and developmental disorders among children aged 2–8 years in rural and urban areas—United States, 2011–2012. *MMWR Surveillance Summaries* 66 (8): 1.

Saunders, Peter. 2003. *Social theory and the urban question*. New York: Routledge.

Smith, Trenton G. 2011. *Obesity and the economics of prevention: Fit not fat*. Oxford: Oxford University Press.

Stanley, Dick. 2003. What do we know about social cohesion: The research perspective of the federal government's social cohesion research network. *Canadian Journal of Sociology/Cahiers canadiens de sociologie* 28: 5–17.

Sundquist, Kristina, Gölin Frank, and Jan Sundquist. 2004. Urbanisation and incidence of psychosis and depression. *The British Journal of Psychiatry* 184 (4): 293–298.

Swinburn, Boyd A., Gary Sacks, Kevin D. Hall, Klim McPherson, Diane T. Finegood, Marjory L. Moodie, and Steven L. Gortmaker. 2011. The global obesity pandemic: Shaped by global drivers and local environments. *The Lancet* 378 (9793): 804–814.

UN Habitat. 2016. *World cities report*. New York: United Nations Human Settlements Programme (UN-Habitat). https://unhabitat.org/books/world-cities-report/.

van den Bosch, Matilda, and Andreas Meyer-Lindenberg. 2019. Environmental exposures and depression: Biological mechanisms and epidemiological evidence. *Annual Review of Public Health* 40: 239–259.

Vlahov, David, and Sandro Galea. 2002. Urbanization, urbanicity, and health. *Journal of Urban Health* 79: S1–S12.

Vlahov, David, Sandro Galea, and Nicholas Freudenberg. 2005. The urban health 'advantage'. *Journal of Urban Health* 82 (1): 1–4.

WHO. 2017. *Obesity and overweight factsheet*. Geneva: World Health Organization. http://www.who.int/mediacentre/factsheets/fs311/en/.

WHO, and UNICEF. 2017. *Progress on drinking water, sanitation, and hygiene*. Geneva: World Health Organization.

Wirth, Louis. 1938. Urbanism as a way of life. *American Journal of Sociology* 44 (1): 1–24.

CHAPTER 4

Nature as a Solution

Abstract This chapter provides an overview of the different ways that nature benefits people, sometimes called ecosystem services. We list the ecosystem services that are key in urban contexts, describing the serviceshed they operate within. We then focus on the evidence that nature exposure benefits health specifically, discussing a few key ecosystem services that are related to mental and physical health. We present the major theories for why exposure to nature reduces stress and provides mental health benefits. Finally, we present a brief history of natural infrastructure in cities and then compare the current amount of natural infrastructure with the amount research says would be ideal in the future.

Keywords Attention restoration theory (ART) • Combined sewer systems • Ecosystem services • Mental health • Serviceshed • Stress reduction theory (SRT)

Nature Can Make the Urban World Humane

The urban lifestyle brings with it some immense benefits to economies and individuals. But as we showed in Chap. 3, life in cities has often been inhumane. While we have solved many aspects of the urban health penalty, obesity and mental health are two areas where there remain some health problems associated with urban living. This chapter argues that nature can be a partial solution, encouraging recreation and alleviating the urban

R. McDonald, T. Beatley, *Biophilic Cities for an Urban Century*,
https://doi.org/10.1007/978-3-030-51665-9_4

psychological penalty. In other words, nature-based solutions can be one way to make our cities more humane, healthier places.

This chapter gets into the details of the relationship between nature and human well-being. We begin by providing an overview of the different ways that nature benefits people, sometimes called ecosystem services. We then focus on the evidence that nature exposure benefits health specifically, discussing a few key ecosystem services that are related to mental and physical health. Finally, we present a brief history of natural infrastructure in cities and then compare the current amount of natural infrastructure with the amount research says would be ideal in the future.

How Does Nature Make Cities More Humane?

Nature provides many benefits to those living in cities that can make life there a little more humane. For instance, the clean drinking water of urban residents depends, in part, on natural vegetation preventing erosion into the city's reservoir. The amount of tree canopy cover in a region affects its air quality. Parks and other urban public space contain natural features like lawns, forests, and lakes that are amenities to people who recreate there. All of these are examples of *ecosystem services*, "the components of nature, directly enjoyed, consumed, or used to yield human well-being" (Boyd and Banzhaf 2006).

Ecosystem services can be provided by specific types of natural features, such as remnant natural habitat, street trees, parks, constructed wetland to mitigate stormwater, or green roofs (McDonald 2015). Collectively, these features are increasingly called "natural infrastructure" or "green infrastructure." In this book, we use "natural infrastructure" in its broad sense, to apply to any of these natural features that provide important ecosystem services. It is important to recognize, however, that different natural features supply different bundles of ecosystem services, so the natural features a city should have to mitigate the urban heat island are different from what are needed to provide mental health benefits. Moreover, there is a continuum of natural infrastructure, from non-anthropogenic (e.g., a wilderness area) to human managed (e.g., critical natural habitat, their ecosystem processes undoubtedly affected in many ways by nearby urban areas) to entirely anthropogenic (e.g., a constructed wetland). From the perspective of this book, regardless of where a piece of natural infrastructure sits along this continuum of naturalness, it is an important natural infrastructure if it is supplying ecosystem service benefits.

An ecosystem service occurs when something an ecosystem does is increasing human well-being. By definition, therefore, ecosystem functions become ecosystem services only when there is someone who is demanding that service (McDonald 2009). For instance, the way a tree's roots can stabilize soil and prevent erosion is the ecosystem function, while the ecosystem service is avoided erosion that provides some economic benefit to people. The distinction between ecosystem function and ecosystem service is important for urban planning. One of the most common errors in planning for ecosystem services is to falsely assume that all features of a particular land-use type generate the same ecosystem function, and are thus identical in ecosystem service value (McDonald 2015). As we shall see, context matters greatly for the value of a service.

Generally, natural infrastructure need to be within a certain distance from people for the ecosystem functions it generates to be useful as an ecosystem service (McDonald 2009). This zone is sometimes called the "serviceshed" (Tallis and Wolny 2011), after the familiar concept of a watershed. The size of this serviceshed varies widely among ecosystem services (Table 4.1), which affects where urban planners and conservation

Table 4.1 Ecosystem services of greatest relevance to urban areas. Also listed is the spatial scale at which they operate

Ecosystem service	*Spatial scale*
Provisioning services	
Agriculture	Regional to global
Water (quantity)	100s km—upstream source watershed
Cultural services	
Aesthetic benefits	10s km—area of daily travel by urbanites
Recreation & tourism	10s km—area of daily travel by urbanites
Physical health	10s km—area of daily travel by urbanites
Mental health	10s km—area of daily travel by urbanites
Spiritual value and sense of place	Varies—often local, but can be up to global
Regulating services	
Water (quality)	100s km—upstream source watershed
Stormwater mitigation	100s m—downstream stormwater system
Mitigating flood risk	100s km—downstream flood-prone areas
Coastal hazard mitigation	10s km—coastal zone
Air purification	100s km—regional airshed
Heat mitigation	<100 m—varies with solar angle

Adapted from McDonald (2015)

practitioners need to protect, restore, or create natural infrastructure. Some services, like temperature reduction from the shade cast by street trees, is very local, operating over the scale of meters. Others, like people's access to parks for day-to-day recreation, operate over the scale of tens of kilometers. Ecosystem services involving water happen within watersheds, which can vary from small to quite big, and are directional: actions upstream affect water quantity and quality downstream, while actions downstream do not affect points upstream. The serviceshed of an ecosystem service is not a simple function of Euclidean distance but is determined by the physics of the ecosystem service in question.

Why are ecosystem services so often degraded over time in cities? Most ecosystem services that cities depend on are public or common goods. These are goods that economists define as non-excludable, meaning that people have access to the goods irrespective of whether they have paid for them or not. A classic finding of environmental economics is that public and common goods suffer from market failure. This occurs because there is no financial incentive linking the use of a resource with the cost of providing the resource, so there is a tendency for free markets to underprovide public or common goods. Therefore, we can expect from economic theory that adequate natural infrastructure for ecosystem service provision will not be maintained by the action of the market. The available empirical data back up this observation. For instance, urban parks for recreation, generally a common good (most urban parks do not have entry fees), will tend not to be provided by private land developers, at least at the level of provision society would prefer. Therefore, there is strong justification for government intervention to ensure adequate natural infrastructure for ecosystem service provision (McDonald 2015).

Categories of Ecosystem Services

Ecosystem services have traditionally been divided into categories (Table 4.1), following those categories used in the landmark Millennium Ecosystem Assessment (2003). The first category of ecosystem services is provisioning services, the products people obtain from ecosystems and directly consume. One clear example of a provisioning service is crop production. Two other examples would be livestock production and aquaculture. Because many provisioning services create products that can be transported over long distances, provision often occurs in rural areas, with their products transported to the city center. Perhaps the most important provisioning services for cities is providing a sufficient quantity of water to their residents, who need water for drinking, sanitation, cleaning, and much more.

Another category of ecosystem service is regulating services, defined as the benefits people obtain from the regulation of ecosystem function. For instance, in riparian systems, natural floodplains allow flood waters to spread out, lessening peak flows and reducing flooding risk in downstream urban areas. Similarly, wetland, oyster reefs, mangroves, and coral reefs may mitigate the risk of flooding to cities during storms. Tree cover in cities can reduce air temperature by providing shade and transpiring water, which improves human health during heat wave events. Since these "risk-reduction" services (a subset of all regulating services) have human health benefits, we will describe them in more detail.

Finally, the Millennium Ecosystem Assessment (2003) recognized another category, cultural services, the nonmaterial benefits people obtain from ecosystems. For example, natural infrastructure can be very important to urban dwellers aesthetically. Whether it is street trees, urban parks, or a view of beautiful mountains, an attractive view has a tangible, quantifiable value to the well-being of those in cities. Recreation opportunities for urban residents are another important benefit of urban natural areas, as is the potential value of natural areas as tourist destinations. Perhaps most importantly for the purpose of this book, there are important health benefits urban residents obtain by being near natural areas, including reductions in obesity and increases in mental health. These health-related ecosystem services are described in more detail in the next section.

Health and Nature Exposure

In this section we examine the small subset of ecosystem services that directly affect health and can thus potentially play a role in helping counteract the "urban health penalty." There is expanding evidence that exposure to nature can be good for health. However, urban nature's relationship to health has been shown to be complex, operating through numerous pathways (Hartig et al. 2014; Hartig and Kahn 2016), with some significant research gaps remaining in current scientific understanding (Frumkin et al. 2017). In this section, we present first the ecosystem services that relate to physical health, then those that relate to mental health. Physical and mental health are of course linked, and improvements in one often lead to improvements in the other. Nevertheless, it is helpful to consider

Fig. 4.1 Forested hills close to downtown Taipei, Taiwan. Urban parks like this one have multiple benefits for physical and mental health. (Photo credit: Ken Marshall (public domain))

each in turn to discuss the distinct pathways by which nature provides benefits (Fig. 4.1).

Note that this is an emerging field of science, and the quality of evidence varies by ecosystem service. For instance, most studies directly looking at the relationship between nature and health use observational rather than experimental data, controlling for other potentially confounding variables through statistical methods (McDonald et al. 2018). This literature also often uses crude proxies of nature exposure from remote sensing, like the normalized difference of vegetation index (NDVI) within some buffer zone around a person's home. By contrast, for some of the regulating services which have a clear physical variable of interest, the science is quite more advanced. Regulating services often have a well-understood logic chain relating the ecosystem function (e.g., shade from a tree) to a change in physical variable (e.g., temperature) to a change in risk to human health (e.g., reduced mortality and morbidity during heat waves).

Physical Health

One major way in which parks and other natural features affect health is by encouraging physical activity during recreation. For instance, a study in Los Angeles by Wolch et al. (2011) tracked the health of more than 3000 children across two decades. They found that children living within 500 meters of a park had lower body mass index (BMI) and better health outcomes. At a broader scale, Ewing et al. (2003) reviewed all United States counties and correlated an index of sprawl with reported walking data and measures of BMI. They found that counties with more sprawl had less walking activity and higher BMI. Research continues into the mechanisms by which urban form affects physical activity. Hartig et al. (2014) described three activity domains: our lifestyle while at work and school, which is little influenced by proximity to parks; transport, where the availability of bike lanes and walkable streets can encourage physical activity while commuting; and leisure activities, where park availability may significantly increase physical activity during recreation.

Increasing access to parks is now seen as one part of an overall strategy to create a more active community. For instance, increasing access to parks is now one of the strategies listed in Centers for Disease Control guidelines to increase physical activity in a community (CDC 2011). The guidelines suggest that as important as the amount of parkland within walking distance of residents is ensuring there are features (playgrounds, walking trails) that residents want to use. While nature and greenspace in the park can be helpful in making a park appealing, it is just one part of having a usable park. Similarly, another strategy listed in the CDC guidelines is creating walkable streets. Increasing street canopy cover can be one part of making street more walkable, but there are many other dimensions of walkability, such as ensuring pedestrian safety and ease of movement.

Flood Risk Reduction

Cities are increasingly using natural infrastructure to reduce risks by providing important water-related ecosystem services, like stormwater mitigation (McDonald 2015). Wetlands and constructed natural features like bioswales can slow the flow of stormwater and increase infiltration into the soil. Natural infrastructure also acts as a filter, reducing concentrations of some pollutants. There are numerous specific types of best management practices for stormwater mitigation, which go by varying names such as bioswales, bioretention cells, rain gardens, and permeable pavement. There is a large evidence base from site-level monitoring that shows that natural infrastructure can achieve its goals of stormwater mitigation (NRDC 2012), although many cities struggle to construct enough natural infrastructure at a large enough scale to make a meaningful difference to watershed hydrology (Pennino et al. 2016). Another concern of many municipalities installing natural infrastructure is the overall lifetime cost of maintenance of the infrastructure, which needs to be periodically cleaned out.

When cities have combined sewers, stormwater mitigation can also help avoid combined sewer overflows, and thus prevent untreated sewage from flowing into area waterways. For example, in Washington, DC, the city's stormwater utility has put in place what is in effect an impervious surface fee on existing developments (McDonald 2015). This creates an incentive for landowners to reduce impervious surface cover or install natural infrastructure. Under Washington's system, new development is held to a higher stormwater retention standard and must capture a certain volume of rainwater on site. If a developer cannot cost-effectively meet the stormwater retention standard on site, they can alternatively choose to purchase stormwater retention credits from someone who has captured an equivalent amount of stormwater on their property.

Another related water-based type of nature-based solution is using natural infrastructure to reduce the risk of floods in floodplains. While maintaining natural vegetation in the watershed can reduce peak flow in a river, for many floodplains, there will always be periods in which the river rises over its banks. A more crucial ecosystem service therefore is the ability of a floodplain to convey and store water. A floodplain covered in natural habitat can perform this service, but so too can any other land-use that can withstand occasional flooding (Opperman et al. 2009). Two common compatible land-uses in floodplains are farming and ranching. While these activities are disrupted by floods—crops may be lost, livestock may have to be moved temporarily—the economic impact is generally modest compared with the damage to buildings and other hard infrastructure.

One famous example of this strategy is the Room for the River program in the Netherlands (Room for the River 2014). The Rhine flows more than 1000 km though some of the major cities in Europe before draining into a large delta in the Netherlands. There have been problems for centuries with flooding in the Rhine. In response, the Dutch have massively engineered the Rhine, building high levees along its length and changing the course it takes to get to the sea. However, damages from flooding have continued to increase. Massive floods in 1993 and 1995 caused significant damage in the delta. The Dutch decided to reverse course and created a new policy called Room for the River. The levees would be removed from the Rhine, creating space for the river to flood. This retreat means that in a few places, some infrastructure, like buildings and roads, will have to be removed from the floodplain.

Cooler and Cleaner Air

Natural infrastructure can also help cool and clean the air. Trees and other vegetation can shade impervious surfaces, limiting their ability to absorb solar energy from the sun and heat up. Plants also transpire water as they grow, which cools the air just as you are cooled when sweat evaporates off your body. Shade and transpiration can reduce air temperatures in cities on hot summer days, which can have real health benefits. In most years, heat waves are the largest weather-related source of mortality globally, killing on average 12,000 people (McMichael et al. 2004), so any reduction in temperature that can reduce this mortality would be an important victory for public health. One study by Anderson and Bell (2011) found that each 0.6 °C (1 °F) increase in air temperature increased the relative risk of mortality in US cities by 2.5%.

Cities are increasingly recognizing heat waves as a growing public health threat and creating heat action plans, which list ways to reduce the risk to their citizens. For instance, Philadelphia recently created a plan to help communities during heat waves (Jaramillo and Conde 2019). The plan proposes the creation of cooling centers, as well as increasing the hours to community pools, so residents without air conditioners have free options that can help cool them off. There are also outreach programs to help ensure elderly residents know how they can get help. As part of this broader heat action plan, Philadelphia is also proposing increased tree planting along streets and in parks to create cooler outdoor conditions. This balance of tree planting with a mix of other public health solutions is increasingly common in urban heat action planning.

Trees and other vegetation not only cool the air; they also help clean it (McDonald et al. 2016). Leaves serve as surfaces for particulate matter to dry deposit on, just as dust settles out on furniture in your house. This is important because particulate matter (PM) is a major public health threat, with outdoor PM killing more than three million people a year. PM, when inhaled into the lungs, causes serious health impacts including cardiovascular disease and asthma. Trees can also modulate concentrations of ozone and other atmospheric pollutants as well. However, the relationship of trees to air quality is complicated, as trees also alter the flow of air, sometimes trapping dirty air from automobiles in places where pedestrians are walking. Vegetation also emits volatile organic compounds, which can locally increase air pollution by encouraging the formation of more PM and ozone (Kroeger et al. 2014). Guidelines exist for tree planting that minimize the risk of these negative consequences, so cities should not be afraid of planting trees, merely thoughtful about their siting and composition.

Cities are increasingly planting trees in part for their ability to improve air quality. One interesting example is in Louisville, Kentucky, in a research study called Green Heart. This study, originated by the University of Louisville Medical School in partnership with The Nature Conservancy and others, is the first neighborhood-scale controlled public health experiment on the links between nature and health. In an intervention neighborhood, urban tree cover is being significantly increased, while in a nearby control neighborhood, tree cover is being left undisturbed. Air quality, temperature, and health outcomes are being tracked in both neighborhoods over at least five years. The hypothesis that is being tested is that there will be an increase in air quality and a decrease in air temperature, which will lead to better health outcomes for the residents of the intervention neighborhood.

Mental Health

Spending time in the natural environment or interacting with natural features (street trees, parks, and gardens) not only is of importance for physical health but also has benefits for mental health. The strongest evidence for a nature/health relationship relates to stress. Nature exposure has been shown to reduce stress, whether measured through self-reporting (e.g., Stigsdotter et al. 2010) or from levels of cortisol (e.g., Thompson et al. 2012). This effect has also been shown in workplaces, with greater indoor contact with plants correlated with less job stress, fewer subjective health complaints, and fewer sick days (Bjørnstad et al. 2016). Exposure to nature has also been shown to be correlated to an improvement in self-reported mental health (Van den Berg et al. 2015), as well as a reduction in the incidence of depression and schizophrenia (Sundquist et al. 2004), although literature is less definitive than for stress reduction. In the discussion that follows, we first discuss reasons why nature exposure might reduce stress, then turn to other potential mechanisms by which nature exposure might affect mental health more broadly.

There are clearly multiple pathways by which interaction with nature can reduce stress (Bratman et al. 2012; Van den Berg et al. 2015). One simple but important pathway is that natural areas may simply have less environmental stressors. For instance, as discussed in Chap. 3, noise pollution increases stress levels. To the extent that noise pollution in parks or natural area is less, it is logical that time in these quieter environments might lower stress levels. Moreover, vegetation like street trees can also serve to muffle noises from traffic, which could have stress reduction benefits for urban benefits.

However, many researchers have suggested that there is something about nature itself, above and beyond the diminution of environmental stressors, that reduces stress. There are two main theories of how this stress reduction occurs, which are not mutually exclusive.

One theory is called attention restoration theory (ART), first proposed by Kaplan and Stephen Kaplan (1989). This theory argues that modern life takes a lot of directed attention—judging the intention of passing strangers, monitoring the sounds coming from our smart phones, answering the flood of work emails. All this directed attention is tiring for our psyches, which did not evolve for such sustained directed attention. A natural setting, by contract, requires less directed attention, which means a visit to a natural setting is a restorative break. Some studies have suggested that natural settings draw our attention to them in a pleasant, effortless manner, while allowing our minds to wander. This "soft fascination" has been confirmed with experiments where people's eye movements were studied (Berto 2005). Some studies have suggested that the intermediate level of visual complexity in natural settings, where fractal patterns recur at many scales, is key to the ability to create a restorative break of "soft fascination" (Van den Berg et al. 2016). After the break, our capacity to give directed attention is increased again. For example, after a visit to a natural setting, scores on tests that require directed attention and short-term memory go up significantly (Berman et al. 2008).

Another theory is called stress reduction theory (SRT). As discussed in Chap. 3, cities are inherently stressful environments for a variety of reasons. As a theory SRT posits that there is something uniquely soothing about natural environments, beyond the mere reduction in environmental stressors or demand for attention, due to our having evolved as a species for these kinds of natural environments. Natural settings thus provide a break from the stress, above and beyond what would happen in other similarly quiet and less stressful places. Studies have shown for instance that stress hormones go down after a walk in the woods (Ulrich 1983), which is some evidence for this theory.

There are other pathways by which nature can increase health (McDonald 2015). For example, social cohesion, the willingness of members of a society to cooperate with each other in order to survive and prosper (Stanley 2003), may be facilitated by having greenspace. One study (Maas et al. 2009) reported that access to urban parks and greenspace contributed to a unique sense of place and increased social cohesion. Another (de Vries et al. 2013) showed a correlation between street-level greenery and perceived social cohesion in the neighborhood. One of the most famous papers on this topic took place in Chicago (Kuo et al. 1998). Public housing residents were randomly assigned to apartments by the housing authority, with some apartments having nearby public greenspace and others without. Those living with nearby greenspace had greater social contact with their neighbors than those residents without.

Finally, we focus in this book on another mechanism by which nature exposure can affect health: biophilia. According to the biophilia hypothesis, humans' need for nature here goes beyond mere reduction of the urban health penalty but is part of the overall need of the human spirit. The biophilia hypothesis is discussed extensively in Chap. 5.

Regardless of the pathways by which natural infrastructure affects mental health, many cities have begun programs to create natural infrastructure to improve the mental health of their residents. One interesting example is the Nature Sacred program. This program puts an emphasis on creating public greenspaces, often small pocket parks close to low-income urban neighborhoods, schools, hospitals, or prisons. They try to co-design with the local community parks that use natural features to create a healing or "sacred" space. The Nature Sacred program has created around 130 places, concentrated in the mid-Atlantic of the United States. There are many other examples of programs like this that have made the healing power of nature central to the justification for new parks.

A Brief History of Natural Infrastructure in Cities

In one sense, cities have always depended on nature for provisioning ecosystem services, like food, water, and wood. However, urban areas have often been conceptualized as the opposite of rural or natural areas, perhaps depending on a broader hinterland but also a special place of human creation that is the antithesis of nature. The growth patterns of traditional cities often left little room for natural features. For instance, reviews of historical maps for European cities show that through the fifteenth century, cities had almost no street trees or public greenspace. Later, I briefly summarize how we got from then to now, but refer readers interested to Lawrence's (2008) detailed history of urban nature.

Consider the invention of the street tree. Street trees now seem like a commonplace idea, but it is important to realize that they were an innovation, which rose in popularity at a specific moment in time. One could say, with only a little exaggeration, that the Dutch invented the street tree. Trees traditionally were used to stabilize canal banks. In cities in the Low Country, these canals came into the city center. The merchant class who owned houses began to advocate for planting trees along other public streets, for their beauty and shade.

Another movement, perhaps best exemplified in France, played a role in the rise of street trees. In royal gardens and along major roads, there was the development of the idea of an *allée*, a long ceremonial path or road lined with trees. This visual feature was useful for displays of marching troops and provided a place for pleasant strolls or carriage rides for the royal court. Over time, the French (and other countries) began importing this idea of an *allée* into the city. Often these were in private gardens such as the Jardin de Tuileries in Paris, reserved originally for royalty.

Medieval cities did have open greenspaces, but they were predominately for the wealthy or for the church (e.g., graveyards). Beginning in seventeenth century, a movement began to open these greenspaces to the public. This movement related to the slow rise of democracy and slow decline of power of the monarchy. Sometimes royal spaces were opened to the public, for example, the Jardin de Tuileries, which was created as a private garden in 1564, was opened to the respectably dressed members of the public in the seventeenth century, and became a fully public park (the National Garden) in 1794 after the French Revolution overthrew the monarchy. By the nineteenth century, it was quite common for new urban greenspaces to be included when new neighborhoods were designated.

For example, the central part of Manhattan in New York City was set aside for Central Park, whose construction began in 1857.

Take a moment to consider how much our urban landscapes have changed in the past four centuries (Lawrence 2008). We have gone from essentially no urban street trees to cities, at least those in appropriate climates, commonly having 20% or more urban tree canopy cover. Similarly, we have gone from essentially no public greenspace in cities to having internationally agreed-upon goals that all urban dwellers globally should have equitable access to a park within walking distance (Sustainable Development Goal 11.7). Humans have invented new types of urban nature over the centuries. While the spread of these innovations was sometimes slow, it was inexorable, and now most cities consider street trees and parks an essential part of their offering to their citizens.

This innovation in new forms of urban nature took place without the benefit of modern science, which has quantified in greater precision than was previously possible in the ecosystem service benefits of natural infrastructure in cities. While citizens in the Low Country could understand that having street trees increased their quality of life, they could not quantify their benefits toward reducing air temperatures, managing stormwater, or increasing mental health. We now know much more about the health benefits of having urban nature in our cities than ever before. It is worth asking, do our cities have enough nature to meet our health needs?

Nature Exposure in Cities Currently

To answer the aforementioned question, we must be able to say something about how much nature in cities is sufficient to meet our health needs. While this is a nascent area of study, several studies have begun to try to measure the dose-response curve, understanding how much exposure to nature is needed to receive benefits to health. For example, Lovasi et al. (2008) found that a one standard deviation increase in tree density (an increase of 343 trees per square kilometer) near homes was associated with a 29% reduction in the prevalence of asthma. A study in Brisbane City, Australia, found that those who visited outdoor greenspaces for 30 minutes or more per week had a 7% reduction in incidences of depression and 9% reduction in incidences of high-blood pressure (Shanahan et al. 2016). More recently, a study in southern England by Cox et al. (2017) tested how neighborhood tree cover within 250 m of a house correlated with the prevalence of several mental health diseases. There was a threshold response, with 50% less depression and 43% less stress in neighborhood with more than 20% forest cover and 56% less anxiety in neighborhoods with more than 30% forest cover.

Of course, nature exposure cannot be collapsed just into one measurement like time in a park or percent greenspace around a home. One review study by Francis et al. (2012) suggested that the quality of natural areas is more important for the magnitude of mental health benefits than is the quantity. That is, it is more important for your mental health to have certain key characteristics present in the greenspace nearby, rather than just more greenspace. One key characteristic is a quieter part of the park, where reflection and solitude are possible. It is also important that all populations, and especially those vulnerable to mental health issues, have easy access to the park and feel safe using it. The magnitude of the impact of nature on physical and mental health likely varies by sociocultural context and depends on complex interactions between urban form, human society, and natural features (Hartig et al. 2014).

Still other studies have looked at whether the biodiversity of plants and animals affects the magnitude of the mental health benefits. Fuller et al. (2007) study did find an association, with more biodiverse parks correlating with greater mental health benefits. However, there are potentially confounding factors, such as the fact that large parcels often tend to have greater biodiversity. Large parcels are, of course, unique for other reasons as well. For instance, the chance for solitude may increase in a large parcel.

Because the literature on dose-response for nature exposure and mental health is so nascent, any extrapolation from the existing literature must be considered preliminary. However, one can use this existing literature to ask, what fraction of the world's urbanites get enough nature now? To address this question, one study (McDonald et al. 2018) used forest cover data for 245 major cities globally (taken from McDonald et al. 2016), and classified 1 km^2 neighborhoods into the same forest cover categories used by Cox et al. (2017). This study estimated that only 13% of urban dwellers are living in neighborhoods with more than 20% forest cover, the threshold found by Cox et al. (2017) to be needed to receive a protective effect against depression, stress, and anxiety. This result suggests that most urban dwellers are living in environments with low levels of nature exposure and may thus be at greater risk of the urban psychological penalty.

Trade-Off Between Density and Natural Infrastructure

Astute readers may notice a potential trade-off here. On the one hand, density is part of what makes cities our greatest invention, as discussed in Chap. 2. More density drives more interaction, and interaction is what drives many of the economic and social benefits of agglomeration. Denser cities also have lower energy and resource use due to economies of scale. For similar reasons, studies that have focused on the size of cities have found that larger cities have greater economic productivity, patent generation, and economic diversity (Bettencourt et al. 2007). Of course, larger cities also tend to be denser, so the two measures are correlated. For the purposes of this section, we will focus on density rather than on sheer size of city, although similar arguments could be made for city size to the extent it is correlated with density.

On the other hand, urban density may make it hard to have a lot of natural features in those dense urban neighborhoods. Certainly, the percent of land in greenspace in a neighborhood is negatively correlated with density. Urban tree canopy cover is also negatively correlated with density. It should be stressed that these correlations are weak and are not absolute barriers. But they reflect a real reality that it is easier to have adequate provision of natural infrastructure of the kind recommended by Cox et al. (2017), who found that >30% forest cover was needed for full mental health benefits, at suburban densities rather than at the densities of an urban core like Manhattan.

One interesting question, not yet fully explored in the literature, is whether there is some happy middle ground. Is there an intermediate density at which there are many of the benefits of compact cities but still adequate provision of natural infrastructure? While this question has not been fully answered in the urban planning and conservation literature, it might be possible. For instance, a series of transportation studies, most notably by Newman and Kenworthy (1989), have related urban population density to transport-related energy consumption per capita. There is a negative relationship, with greater urban population density being correlated with greater use of mass transit and less use of personal automobiles, and hence lower overall transport-related energy consumption per capita. Cities with densities above 50 people/ha, such as many European cities, have relatively low transport-related energy consumption compared to American or Australian cities (Lefèvre 2009). This level of density can be achieved with two- to three-story row-houses, or with three-story garden apartment buildings (e.g., Campoli and MacLean 2007). There are neighborhoods at this density that also have significant public open space and urban tree canopy cover.

Note also that there are some types of natural feature which don't necessarily take up space from housing and other urban uses. For instance, urban street trees are often fit into the public right of way between the sidewalk and the street and are feasible to include even in very dense urban neighborhoods like Manhattan, which has an average density of 273 people/ha (World Population Review 2020). Because urban tree canopy cover does not exclusively take space from roads and sidewalks, but rather covers them, an increase in urban tree canopy does not decrease the space available for roads or sidewalks. Similarly, green roofs are increasing in popularity, and the technology for their design and upkeep is improving. Since green roofs turn the roof of a building into a piece of natural infrastructure, rather than requiring space, there is the potential for even dense urban neighborhoods to increase their natural infrastructure using green roofs.

Closing

This chapter has shown that natural features in cities like remnant nature habitat, parks, and street trees can increase human health and well-being. We believe the empirical evidence suggests that natural features can reduce the remaining urban health penalties of obesity and mental health. The data suggest that nature exposure can alleviate but not eliminate this urban health penalty, while also providing other ecosystem service benefits to human well-being. While nature is not the sole solution to the urban healthy penalty, it is one of the potential solutions cities can turn to. However, despite the increasing scientific knowledge of these benefits, available data on urban forest cover for cities globally suggest most cities do not have adequate nature for their citizens. Thus, while urban nature is providing significant benefits to health, cities could and should increase urban nature even more to further increase benefits.

We suggest that nature in cities could be a way for humanity to have our cake and eat it too. We can have all the benefits of an urban, connected, dense world. But we can at the same time live in natureful, humane cities that have improved mental and physical health outcomes for all their residents, rich and poor. Humanity have made some strides over the last few centuries in creating that kind of urban world, as we have refined our innovations of the street tree and the public park. More work remains, of course, to continue expanding urban canopy cover and park access.

However, we believe that the urban century will require a new kind of urban natural infrastructure. The newfound knowledge about nature and its health benefits is providing new insights for urban planners and designers. At the same time, technological advancement is driving changes in how urbanites work, commute, shop, and socialize, all of which are driving further changes in urban form and function. Just as our conception of the value of urban nature is very different than medieval cities, there is no reason that our twenty-first-century digital city can't have a different conception of what nature can be. Nature, far from being irrelevant in the urban century, appears necessary for the urban century's success.

References

Anderson, G. Brooke, and Michelle L. Bell. 2011. Heat waves in the United States: Mortality risk during heat waves and effect modification by heat wave characteristics in 43 US communities. *Environmental Health Perspectives* 119 (2): 210.

Berman, M.G., J. Jonides, and S. Kaplan. 2008. The cognitive benefits of interacting with nature. *Psychological Science* 19: 1207–1212.

Berto, R. 2005. Exposure to restorative environments helps restore attentional capacity. *Journal of Environmental Psychology* 25 (3): 249–259.

Bettencourt, L., J. Lobo, D. Helbing, C. Kuhnert, and G. West. 2007. Growth, innovation, scaling, and the pace of life in cities. *Proceedings of the National Academy of Science* 104 (17): 7301–7306.

Bjørnstad, Siv, Grete G. Patil, and Ruth K. Raanaas. 2016. Nature contact and organizational support during office working hours: Benefits relating to stress reduction, subjective health complaints, and sick leave. *Work* 53 (1): 9–20.

Boyd, J., and S. Banzhaf. 2006. *What are ecosystem services? The need for standardized accounting units.* Washington, DC: Resources for the Future.

Bratman, G.N., J. Hamilton, and G. Daily. 2012. The impacts of nature experience on human cognitive function and mental health. *Annals of the New York Academy of Sciences* 1249: 118–136.

Campoli, Julie, and Alex S. MacLean. 2007. *Visualizing density.* Lincoln Institute of Land Policy.

CDC. 2011. *Strategies to prevent obesity and other chronic diseases: The CDC guide to strategies to increase physical activity in the community.* Atlanta: U.S. Department of Health and Human Services, Centers for Disease Control and Prevention.

Cox, Daniel T.C., Danielle F. Shanahan, Hannah L. Hudson, Kate E. Plummer, Gavin M. Siriwardena, Richard A. Fuller, Karen Anderson, Steven Hancock, and Kevin J. Gaston. 2017. Doses of neighborhood nature: The benefits for mental health of living with nature. *BioScience* 67 (2): 147–155.

de Vries, Sjerp, Sonja M.E. van Dillen, Peter P. Groenewegen, and Peter Spreeuwenberg. 2013. Streetscape greenery and health: Stress, social cohesion and physical activity as mediators. *Social Science & Medicine* 94: 26–33.

Ewing, R., T. Schmid, R. Killingsworth, A. Zlot, and S. Raudenbush. 2003. Relationship between urban sprawl and physical activity, obesity, and morbidity. *American Journal of Health Promotion* 18 (1): 47–57.

Francis, J., L. Wood, M. Knuiman, and B. Giles-Corti. 2012. Quality or quantity? Exploring the relationship between Public Open Space attributes and mental health in Perth, Western Australia. *Social Science and Medicine* 74: 1570–1577.

Frumkin, Howard, Gregory N. Bratman, Sara Jo Breslow, Bobby Cochran, Peter H. Kahn Jr., Joshua J. Lawler, Phillip S. Levin, Pooja S. Tandon, Usha Varanasi, and Kathleen L. Wolf. 2017. Nature contact and human health: A research agenda. *Environmental Health Perspectives* 125 (7): 075001.

Fuller, R., K. Irvine, P. Devine-Wright, P. Warren, and K.J. Gaston. 2007. Psychological benefits of greenspace increase with biodiversity. *Biology Letters* 3 (4): 390–394.

Hartig, Terry, and Peter H. Kahn. 2016. Living in cities, naturally. *Science* 352 (6288): 938–940.

Hartig, Terry, Richard Mitchell, Sjerp De Vries, and Howard Frumkin. 2014. Nature and health. *Annual Review of Public Health* 35: 207–228.

Jaramillo, C., and X. Conde. 2019. Philly debuts community-driven plan to beat rising temperatures. *Plan Philly.*

Kaplan, R., and S. Kaplan. 1989. *The experience of nature: A psychological perspective.* New York: Cambridge University Press.

Kroeger, T., F.J. Escobedo, J. Hernandez, S. Varela, S. Delphin, J. Fisher, and J. Waldron. 2014. Reforestation as a novel abatement and compliance measure for ground-level ozone. *Proceedings of the National Academy of Science* 111 (40): E4204–E4213. https://doi.org/10.1073/pnas.1409785111.

Kuo, Frances E., William C. Sullivan, Rebekah Levine Coley, and Liesette Brunson. 1998. Fertile ground for community: Inner-city neighborhood common spaces. *American Journal of Community Psychology* 26 (6): 823–851.

Lawrence, Henry W. 2008. *City trees: A historical geography from the renaissance through the nineteenth century.* University of Virginia Press.

Lefèvre, Benoit. 2009. Urban Transport Energy Consumption: Determinants and Strategies for its Reduction.. An analysis of the literature. *SAPI EN. S. Surveys and Perspectives Integrating Environment and Society* 2: 3.

Lovasi, Gina Schellenbaum, James W. Quinn, Kathryn M. Neckerman, Matthew S. Perzanowski, and Andrew Rundle. 2008. Children living in areas with more street trees have lower prevalence of asthma. *Journal of Epidemiology & Community Health* 62 (7): 647–649.

Maas, J., S.M.E. Van Dillen, R.A. Verheik, and P.P. Groenewegen. 2009. Social contacts as a possible mechanism behind the relation between green space and health. *Health Place* 15 (2): 586–595.

McDonald, R.I. 2009. Ecosystem service demand and supply along the urban-to-rural gradient. *Journal of Conservation Planning* 5: 1–14.

———. 2015. *Conservation for cities: How to plan & build natural infrastructure*. Washington, DC: Island Press.

McDonald, R.I., T. Kroeger, T. Boucher, Longzhu Wang, and R. Salem. 2016. *Planting healthy air: A global analysis of the role of urban trees in addressing particulate matter pollution and extreme heat*. Arlington, VA: The Nature Conservancy.

McDonald, R.I., T. Beatley, and T. Elmqvist. 2018. The green soul of the concrete jungle: The urban century, the urban psychological penalty, and the role of nature. *Sustainable Earth* 1 (1): 3.

McMichael, A., D. Campbell-Lendrum, S. Kovats, S. Edwards, P. Wilkinson, T. Wilson, R. Nicholls, S. Hales, F. Tanser, D. Le Sueur, M. Schlesinger, and N. Andronova. 2004. Global climate change. In *Comparative quantification of health risks: Global and regional burden of disease attributable to selected major risk factors*, ed. Majid Ezzati, Alan D. Lopez, A. Rodgers, and C. Murray. Geneva: World Health Organization.

MEA. 2003. *Ecosystems and human well-being: A framework for assessment*. Washington, DC: Island Press.

Newman, P., and J. Kenworthy. 1989. *Cities and automobile dependence: An international sourcebook*. Brookfield, VT: Gower Publishing.

NRDC. 2012. *Financing stormwater retrofits in Philadelphia and beyond*. New York: National Resources Defense Council.

Opperman, J., G. Galloway, J. Fargione, J. Mount, B. Richter, and S. Secchi. 2009. Sustainable floodplains through large-scale reconnection to rivers. *Science* 326: 1487–1488.

Pennino, Michael J., Rob I. McDonald, and Peter R. Jaffe. 2016. Watershed-scale impacts of stormwater green infrastructure on hydrology, nutrient fluxes, and combined sewer overflows in the mid-Atlantic region. *Science of the Total Environment* 565: 1044–1053.

Room for the River. 2014. Riumte voor der ivier (English). The Netherlands. Accessed 1 June 2014. http://www.ruimtevoorderivier.nl/english/.

Shanahan, Danielle F., Robert Bush, Kevin J. Gaston, Brenda B. Lin, Julie Dean, Elizabeth Barber, and Richard A. Fuller. 2016. Health benefits from nature experiences depend on dose. *Scientific Reports* 6: 28551.

Stanley, Dick. 2003. What do we know about social cohesion: The research perspective of the federal government's social cohesion research network. *Canadian Journal of Sociology/Cahiers canadiens de sociologie* 28: 5–17.

Stigsdotter, Ulrika K., Ola Ekholm, Jasper Schipperijn, Mette Toftager, Finn Kamper-Jørgensen, and Thomas B. Randrup. 2010. Health promoting outdoor environments—Associations between green space, and health, health-related quality of life and stress based on a Danish national representative survey. *Scandinavian Journal of Social Medicine* 38 (4): 411–417.

Sundquist, Kristina, Gölin Frank, and J.A.N. Sundquist. 2004. Urbanisation and incidence of psychosis and depression. *The British Journal of Psychiatry* 184 (4): 293–298.

Tallis, H., and S. Wolny. 2011. Including people in the mitigation hierarchy: Mapping ecosystem service winners and losers in Colombia. Paper read at Ecological Society of America, Austin, TX.

Thompson, Catharine Ward, Jenny Roe, Peter Aspinall, Richard Mitchell, Angela Clow, and David Miller. 2012. More green space is linked to less stress in deprived communities: Evidence from salivary cortisol patterns. *Landscape and Urban Planning* 105 (3): 221–229.

Ulrich, R. 1983. Aesthetic and affective response to natural environment. *Human Behavior Environmental Advanced Theory Research* 6: 85–125.

Van den Berg, M., W. Wendel-Vos, Mireille Van Poppel, H. Kemper, W. Van Mechelen, and J. Maas. 2015. Health benefits of green spaces in the living environment: A systematic review of epidemiological studies. *Urban Forestry & Urban Greening* 14 (4): 806–816.

Van den Berg, Agnes E., Yannick Joye, and Sander L. Koole. 2016. Why viewing nature is more fascinating and restorative than viewing buildings: A closer look at perceived complexity. *Urban Forestry & Urban Greening* 20: 397–401.

Wolch, J., M. Jerrett, K. Reynolds, R. McConnell, R. Chang, N. Dahmann, K. Brady, F. Gilliand, J. Su, and K. Berhane. 2011. Childhood obesity and proximity to urban parks and recreational resources: A longitudinal cohort study. *Health & Place* 17: 207–214.

World Population Review. 2020. *Manhattan population 2020*. World Population Review.

Stanley, Dick. 2003. What do we know about social cohesion: The research perspective of the Federal Government's social cohesion research network. *Canadian Journal of Sociology* [illegible]

[illegible]

[illegible]

[illegible]

[illegible]

[illegible]

CHAPTER 5

Biophilic Cities: Vision and Emerging Principles

Abstract This chapter presents one way to plan for using nature as a solution: biophilic design and planning. Biophilic cities represent a new vision of global urbanization that puts nature at the center of design and planning. It calls for a shift from seeing cities as places where there are discrete elements of nature to a more integrative and holistic sense of nature. This vision recognizes as well the multiple benefits of immersive nature in cities. Biophilic cities are resilient and sustainable cities, at once providing many ecological and adaptive services, while also providing essential benefits in the form of mental and physical health. In this chapter we explore this vision in some detail and lay out a series of planning principles to help guide cities.

Keywords Biophilic Cities Network • Green roofs • Living walls • Natureful • Street trees • Wildness

If life in a city poses some potential challenges to human health and well-being, and if nature can be part of the solution, then the next obvious question to urban planners and designers is how they should plan to incorporate nature into cities. The rest of this book focuses on the techniques and practices of biophilic design, arguing that it is the best way to create the urban natural infrastructure our cities will need for the urban century

R. McDonald, T. Beatley, *Biophilic Cities for an Urban Century*,
https://doi.org/10.1007/978-3-030-51665-9_5

to be successful. We believe that biophilic design could help create this new generation of urban natural infrastructure.

The premise of biophilia is that as a species, we have an innate connection to nature. We are happier, healthier, and able to lead more meaningful lives when nature is around us, where we live and work. Biophilic designers and planners argue that the integration of nature into urban design and planning is an important step toward future cities that are uplifting, restorative, beautiful, and designed around a sense of connection with and wonder about the natural world. The chapter will review key principles of biophilic design, drawing heavily from the experience and practices of partner cities participating in the global Biophilic Cities Network.

The Vision of Biophilic Cities

The last decade especially has seen the emergence of a remarkable record of evidence, from the fields of environmental psychology, public health and medicine, and economics, showing the many positive benefits of contact with nature. Much of this evidence has been reviewed in Chap. 4 of this book, but to summarize, it is increasingly clear that the ecosystem services supplied by natural features like trees makes us calmer, less stressed, happier, and more focused. Nature helps us to be better human beings.

A foundational concept of biophilic design is biophilia, the view that we have coevolved with the natural world and that we are innately drawn to nature. EO Wilson has described this as "the innately emotional affiliation of human beings to other living organisms. Innate means hereditary and hence part of ultimate human nature" (Wilson 1984). The central idea is that we have coevolved with nature and developed preferences for the qualities of the natural world that have helped us survive as a species. We are drawn to flowers, climbable trees, water, and other natural elements for the survival benefits they provide, and we have evolved to prefer landscapes with "prospect and refuge" (places where we have expansive landscape views but which also provide security and protection). It is not at all surprising that we see the physical and mental health benefits of nature—that we are drawn to water, trees, birdsong—given this deep evolutionary history as a species and the survival benefits bestowed by these natural features (Wilson 1984; Kellert et al. 2011; Beatley 2011).

The emerging vision of biophilic cities builds centrally on these insights about the power of nature. Contact with nature, it is increasingly understood, is an essential element of a healthy, flourishing human life. A key premise is that nature in cities in not something optional, but is essential to leading a happy, healthy, and meaningful life. Along with new research and appreciation of the many ecosystem services provided by nature (e.g., how trees help shade and cool cities), the growing scientific literature shows that nature can no longer be seen as peripheral to city design, but must be viewed as primary and central.

There has been over the last two decades the emergence of a robust practice around biophilic design—designing buildings and increasingly the interior spaces of offices, schools, and hospitals to bring nature into them (sunlight, fresh air, greenery of all sorts). The recognition of the importance of a parallel set of ideas at city scale has led to a biophilic cities movement that believes that we must reimagine cities as places of nature. It extends biophilic design and planning beyond the building to the spaces between and beyond the buildings, to encompass urban neighborhoods and districts, to the larger scale of the city and region, and even connects to efforts to protect natural systems and biodiversity at larger continental and global scales.

A Global Movement and an Emerging Network

We are witnessing a growing global movement in support of naturalizing or re-naturalizing cities, restoring and growing nature everywhere we can in cities, utilizing the special powers of nature as a key foundation for urban design and planning in cities around the world.

It is remarkable how many different organizations, people, and cities are coming to similar conclusions about the essential role of nature in the modern era of cities and working to advance this broad goal. There are lots of other names and language that are being used to advance a nature and cities agenda. Sometimes the term "green cities" is used, though it is often invoked to refer to an even broader vision of cities (e.g., sustainable and ecological), not squarely or directly referring to contact with nature. Richard Louv has used "nature-rich communities" (Louv 2018), and there is an increasing use of "natural infrastructure" or "nature-based solutions" (IUCN n.d.; Connecting Nature n.d.). There is now the concept of a National Parks City, advanced by geographer Daniel Raven-Ellison, who has worked to have London designated as the first (which successfully happened in 2019; National Park City n.d.). Mayors, organizations, and individuals have been encouraged to sign the London

National Park Charter, and many thousands did in the lead-up to the City, officially endorsing the designation. There is now a National Park City Foundation, with its own new network of cities around the world embracing this vision and designation.

There are several other emerging networks supporting or advocating for more nature in cities, and this we believe is a very positive trend. These include the collective blog The Nature of Cities (TNOC), ICLEI's Cities with Nature, and IUCN's Nature-based Solutions initiative, and Cities Connecting Children to Nature (a collaboration between the Children and Nature Network and the National League of Cities), among others (See Table 5.1). We are presently in a positive period of convergence in global thinking about cities, as more individuals and organizations realize nature's central role in reshaping our models of urbanization and urban living.

The Biophilic Cities Network, founded in 2013, grows from a similar strong sense of a need to place nature at the core of design and planning. The Network currently consists of 22 cities that have officially joined as partner cities, as well as thousands of individuals and organizations that have joined by signing the Biophilic Cities Pledge (see Biophilic Cities n.d.-a, b, c). Partner cities include American cities such as San Francisco, Milwaukee, and Norfolk, Virginia, but also cities from around the world,

Table 5.1 The growing importance of the nature-lens on cities. Examples of language and initiatives, in addition to the Biophilic Cities Network

Natural infrastructure
Nature-based solutions (and IUCN's Nature-Based Solutions Initiative) (IUCN n.d.)
Connecting nature (Connecting Nature n.d.)
National park cities (National Park Cities n.d.)
Green and ecological infrastructure (USEPA n.d.)
Natural Capital Cities (Natural Capital Coalition n.d.)
Nature-rich communities or cities (Louv 2018)
Cities connecting children to nature (Children and Nature Network n.d.)
NYC Nature Goals 2050 (NYC Nature Goals 2050 n.d.)
The Nature of Cities (TNOC) Collective Blog (TNOC n.d.)
ICLEI's cities with nature (ICLEI n.d.)
Cities4Forests (WRI n.d.)
European Green Capital Award (EC n.d.)
The Nature Conservancy's Build Healthy Cities Program (TNC n.d.)

including Singapore, Wellington (New Zealand), Edmonton (Canada), and Vitoria-Gasteiz (Spain) (See Beatley 2018).

Much of the focus of the Biophilic Cities Network has been on documenting the stories of innovative practice, and the many different tools and techniques and strategies utilized by cities to foster natureful urban environments. From San Francisco's Bird-Safe Building standards to Portland's Green Streets program to Singapore's efforts at vertical greening, the Network seeks to capture and share these stories, tools, and codes through a variety of means including films, workshops, and a new online Biophilic Cities Journal.

City in a Forest: A Vision of Immersive Urban Nature

The vision of biophilic cities is a holistic one—the view that cities can be places of immersive nature. "Nature" is not a category of thing to be found in a designated place or location in a city—such as a park or a pond or forest—rather, it is a seamless and surrounding condition, one that can be actively grown and regrown, recovered and restored in all the spaces of cities.

Some cities, notably Singapore, have officially changed their mottos to reflect this more immersive vision of cities in nature. Garden City Singapore is now "Singapore in a Garden," a small but immensely important shift in emphasis. And as the city continues to embrace this vision, it increasingly calls itself, Singapore, a Biophilic City in a Garden!

The shift in emphasis is important: we still must overcome the idea that nature can be found in cities but only in certain places—parks, for example—places that require one to travel to or visit. Why not reconceptualize the city as *the* park, as the garden, as the forest? And in fact many American cities are rediscovering and re-invoking earlier mottos and city monikers that reflected this vision (e.g., Atlanta has called itself the City of Forests, Denver's motto has been "City Within a Park").

Nearby Nature and Everyday Wildness

A key premise of the biophilic city is that nature is not something remote and distant, to be visited once or twice on holiday. The experiences of visiting a distant wilderness or national park are important and valuable, but the vision we are putting forth here acknowledges the essential need for nature near to where we live and work. We want and need nature experience daily, if not hourly, and the design and planning of cities to place nature at their center can provide an abundance of opportunities for such experience.

We can best understand the city as a complex, interconnected ecological system; as a place of shared habitat, where life and living systems extend from the microscopic to the continental; as a habitat where biodiversity is everywhere, from subterranean soil to the tops of bridges to the aquatic life of the rivers and oceans. Many biophilic buildings and biophilic designs seek to incorporate new forms of nature, from green rooftops to living walls to skyparks, in turn connecting with ground-level parks, nature, tree canopy, gardens. In this way, biophilic cities are complex hybrids of remnant and human-designed forms of urban nature.

Biophilic Cities Provide Extensive Ecosystem Services

Nature can provide multiple ecosystem services that enhance human health and well-being, as discussed in Chap. 4. Planning for the provision of these ecosystem services is a core part of biophilic design. For instance, urban trees and forests represent one important category of green or ecological infrastructure that biophilic cities invest in. Planting trees and urban forests provide shade and evapotranspiration cooling benefits, as well as providing habitats for birds and other urban wildlife and help with the conservation goals of cities. Many cities have made great strides to protect and expand their urban forests, such as New York City, which reached its one million tree goal. New studies suggest that urban tree planting can have a major impact on urban heat. Efforts are under way in Dallas that serve as a guide to other cities.

The Dallas Urban Heat Island Management Study, completed in 2017, a joint project of the Urban Climate Lab and the Texas Trees Foundation, models the likely impacts of a combination of tree planting and use of cool construction materials. The study found tree planting to be much more effective than other techniques. In the words of the report: "Tree planting and preservation in Dallas can change the weather--producing cooler days and nights than will occur if tree canopy continues to be lost" (Urban Climate Lab 2017, p. 3; see also Goldsmith 2018). The report recommends that at least 250,000 trees be planted in Dallas, in particular places.

Biophilic Cities Are Resilient Cities

Design and planning for biophilic cities can significantly enhance the resilience of a city as well. As adaptation to climate change becomes ever more important, investing in nature makes increasing sense. Green urban elements, from tree planting to rooftop meadows, can be an important response to the expected increase in extreme weather events such as flooding and coastal storms. New coastal parks and greenspaces can provide the needed urban greenspace while also retaining floodwaters and adapting to rising sea level. Increasingly in cities like New York and Norfolk, new public parks are designed to incorporate restored wetlands and designed to be floodable.

The rise of the concept of "sponge cities" in China further shows the ways in which parks and greenspace can respond to recreational and other needs of cities, but also help retain stormwater and moderate local flooding. Norfolk's design for a new resilient park is similarly a recognition that it is possible and desirable to enhance access to nature but to do this in ways that also make the city safer and more resilient.

Biophilic Cities Are Healthy Cities

Biophilic design found some of its first applications in health and medical facilities, a recognition of the essential healing power of nature. Roger Ulrich's groundbreaking research in the 1980s found that patients in hospital rooms with views of nature recovered more quickly and needed fewer pain killers, setting the stage for much future research and helping to convince hospital designers of the importance of enlisting biophilic design in the medical healing and recovery process (Ulrich 1984).

The evidence is strong and growing that designing in nature can generate significant positive health benefits. As discussed in Chap. 4, there are the direct positive health effects from urban nature, such as the cleaner and cooler air from investing in urban trees, but there are many more indirect benefits to health. Greener, more natureful urban neighborhoods in turn propel residents outside and lead to more active, healthier lifestyles. Trees, parks, and other elements of nature also serve to bring people together, help to foster friendships and social relationships, and help to build essential social capital, all delivering important health benefits.

Biophilic cities present the potential for human and natural flourishing to happen together, with an emphasis on restoring ecosystems and biodiversity in cities but also creating new opportunities for residents to connect with and engage this nature around them. Among the essential conditions for human flourishing, we believe, is a deeper sense of purpose and meaning in life, something nature is especially able to provide. Opportunities for hands-on work studying and restoring the nature in cities are an especially potent source of meaning in life.

Repurposing Land, Making Room for Nature

Biophilic cities find new and creative places to grow and insert new nature. This can be a rooftop garden in a high-rise building but can also be on a vacant lot or the median strip of a roadway. Creatively repurposing vacant or leftover land and converting paved surfaces of the city to trees and nature are key goals in biophilic cities.

Many of the cities in the Biophilic Cities Network are in fact doing many creative things to repurpose leftover spaces for nature. There are several impressive examples of partner cities that have done this. Milwaukee has been consolidating vacant lots into new pocket parks in neighborhoods that especially need them and, in examples like Fondy Park, designing them in ways that capture and retain stormwater. Similarly, the City of St. Louis has set citywide targets for and engaging residents around the planting of butterfly gardens on sites throughout the city. The city set an official goal of planting 250 butterfly gardens, but impressively exceeded (there are now more than 400) in this wildly popular initiative.

Sandra Albro in her new book *Vacant to Vibrant* tells the story of her efforts and the work of the Cleveland Botanical Gardens to assist the conversion of vacant land in three legacy cities (Cleveland, Buffalo, and Gary; See Albro 2019). Montreal has famously created a Green Alleys initiative, allowing residents to take over and re-green these spaces between and behind their homes, and other cities from Chicago to San Francisco, have implemented similar green alleys and laneway programs. San Francisco has been pioneering new planning tools and mechanisms, which will be discussed in more detail in Chap. 6.

Biophilic Cities Embrace Wildness in the City

The vision of biophilic cities is one that seeks to introduce new forms of wildness by restoring or "rewilding" the many (formerly) natural places and interstitial spaces in the city. The natural elements of wildness existing in cities will vary, but often include its topography and unique natural features, its hydrology and riparian systems, and urban wildlife. Biophilic cities can actively protect and restore these features, build upon them in their planning, and work to educate about and celebrate them.

Many exemplary efforts are underway. In the capital city of San Jose, Costa Rica, work has begun on a project called *Rutas Naturbanas*—a vision of a 25-kilometer network of walking trails along the two major rivers that flow through the city. Most of the city has turned its back on these rivers, and this is an attempt to rediscover and restore a relationship with this riparian wild area.

The return of urban wildness to cities also often entails the restoration of urban ecosystems. Many of the Biophilic Cities Network partner cities have wonderful and inspiring stories of eco-restoration. Vitoria-Gasteiz (Spain) famously converted its municipal airport back to a functioning and biodiverse wetland, now a RAMSAR site. Many smaller-scale examples exist, for instance, Perth's (Australia) conversion of a sterile, energy-intensive, highly chlorinated urban water feature into an urban wetland, a micro-wilderness in the heart of this city (see Biophilic Cities n.d.-a, b, c).

Conservation in a biophilic city begins at home. There are abundant places where new nature can be grown and restored, including on rooftops and wall facades. Yard farming is a new undertaking for some, and there are many impressive ways that urbanites and homeowners can take direct action to enhance habitat and improve the environment. In Portland, remarkably more than 5000 property owners are participating in a backyard habitat certification program run by the Audubon chapter there (Portland Audubon n.d.). In a biophilic city citizens are themselves passionate and motivated about nature and the need to protect it and connect with it, and much of what is needed are avenues and programs to guide and steer these values.

The Importance of Horizontal and Vertical Nature(s)

Cities of the future will necessarily need to be dense and compact. Long-term global sustainability will require this. These are the urban conditions that will help to reduce energy and resource consumption, and help mitigate greenhouse gas emissions, as well as allow cities to become more walkable and transit-friendly (e.g., Newman and Kenworthy 2015). As discussed in Chap. 4, one way to have density and nature is to look for ways to incorporate nature that does not reduce space for buildings. Accordingly, biophilic cities view every rooftop or building facade as a chance to support living nature.

We have entered a new era of vertical nature and greening. One remarkable example is the Bosco Verticale project in Milan, Italy. Designed by Stefano Boeri, these twin residential towers include some 800 trees (as well as 4500 shrubs and 15,000 perennial plants). Another interesting example is Designer's Walk, a residential tower recently approved in Toronto. The work of architectural firm Brisbin Brook Beynon Architects, it even more firmly integrates trees into the structural design of the building, creating a plug and play system for the trees and creating depressions in the structure's floor plates for the trees to sit. Interestingly, the surrounding neighborhood became major advocates for the projects, seeing the building as a neighborhood enhancement (contrary to the usual NIMBYism that prevails in cases of higher density housing). Indeed Brian Brisbin, lead architect, describes the structure as a "terraced hillside of community trees," and residents of the neighborhood concur with this positive description (Beatley 2019).

Rooftops in the vertical biophilic city become important places for nature and human occupation. Examples of major new rooftop parks seem to abound. The 5.4 acre park on the roof of the Transbay Transit Center in San Francisco is a wonderful case-in-point, designed for water collection and reuse (collecting water used in toilet flushing, estimated to save nearly 3 million gallons of potable water annually) and provision of new public picnicking and civic space for the residents.

Biophilic Cities Take a Whole-of-City Approach

Urban natural features can take many forms: rivers, shorelines, and larger ecosystem elements of a city or metropolitan area that tend to form the ecological structure of a place, but then, trees, gardens, greenery in the spaces between and around homes and buildings, and green rooftops and living facades integrated into the design of the structures themselves. We sometimes say that the vision of biophilic cities is a "whole-of-city" vision that sees the opportunity for growing and inclusion of nature at *every* scale, from "room or rooftop to region or bioregion," and every space in between.

In Chap. 6, we will explore in more detail examples of the biophilic design and planning elements and strategies that can be applied at different geographical scales. There are now abundant and inspiring examples of biophilic buildings that incorporate natural light, greenery, natural materials, and biophilic features such as interior and exterior living walls. But there are important actions that can take place at the district or neighborhood scale, such as the design of parks, pedestrian environments, trees and gardens, that together create the context of nature in which a single building or facility sits. At a broader scale we are concerned with citywide plans, urban design strategies, and codes that create a citywide context in which every neighborhood sits in nature, or "within a garden or forest or park." Ideally, these scales are integrative and reinforcing.

Just as biophilic cities follow a "whole of city" vision, they also embrace a "whole of life," in the sense that it is understood that nature and nature-connectedness in cities should exist at all phases and stages of life, from childhood throughout one's adult life and into the elder years. Nature plays an essential role in creating conditions of health and meaning at each stage. Every school in a biophilic city should be designed to encourage outdoor learning and play and offer abundant opportunities for experiencing nature during the course of the day.

Many positive examples now exist of biophilic schools, including, for instance, the Chattahoochee Hills Charter School, near Atlanta, Georgia (USA), where students come to school each day ready to move around and spend time outside. The physical design of the school is not a single structure but a series of smaller buildings that students move among and between over the day. Equipped with books and outdoor gear, the students spend much of the day in the forest and surrounding nature, which becomes an outdoor lab for science and math. Like many schools today, there is also an emphasis on growing food, and planting and tending the school's gardens is something most students are actively involved in. On the other end of the age-spectrum, biophilic cities seek to provide older residents with abundant opportunities to engage in nature through learning, recreating, and volunteering.

Outside Lies Magic (in Cities)

An important element of the vision of biophilic cities is a rediscovery of the outside world. It is a difficult challenge to get more residents outside, as Americans famously spent 90% or more of the day inside homes and offices (and a considerable portion of day traveling in sealed automobiles). We know that we have a serious problem with unhealthy sedentary lives and lifestyles, and there has been concerted work to design for physical activity.

Outside Lies Magic is the wonderful title of a book by Harvard Professor John Stilgoe, who argues for acute attentiveness to the places and common histories around us (Stilgoe 1998). Attentiveness, curiosity, openness to the nature around us are important qualities in a biophilic city. We need cities that will instill and re-instill a sense of magic—a sense of the wonder and awe and mystery that exists everywhere in cities.

Propelling residents outside, then, becomes an important mission in biophilic cities, and there are many creative ways partner cities are doing this. Partly it is about investing in infrastructure that allow hiking and walking in the city. Singapore, for instance, has developed a network of "Park Connectors," trails and pathways that connect high-density population centers with major parks and greenspaces. This network is now more than 300 kilometers in length (Singapore Nparks n.d.). Other cities, from Oslo to Wellington, have similarly developed wonderful trail networks that serve mobility needs but also allow residents to move through a city off-street and to see and experience nature away from automobile-dominated roads and streets (Fig. 5.1).

Hiking the Biophilic City

Biophilic cities prioritize investments in the infrastructure that enables spending time outside. For instance, San Francisco is in many ways in a league of its own when it comes to trails. In 2019 the Bay Area celebrated the 30th anniversary of the beginning of two parallel (and unparalleled) regional trail networks—one along the shoreline edge of Bay (The Bay Trail) and the other along the ridge tops (the Bay Ridge Trail). Each aspires to a trail length of 500 miles+ and together will eventually provide residents with more than a thousand miles of trails, allowing urban residents unusual access to nearby wildness and nature. In San Francisco, it is possible to experience the outdoors and to relatively easily visit wild nature, not far away but providing wonderful opportunities to camp and hike. Other cities, from Rio de Janeiro to Wellington, NZ, have invested similarly in urban and regional trail networks.

Fig. 5.1 An example of a Singapore Park Connector, allowing for pedestrian movement and interaction with nature. (Photo credit: Timothy Beatley)

Biophilic cities, then, are hiking cities. Investment in trees, gardens, and nature of all kinds, even the green elements of biophilic buildings add interest in walking and hiking through the city.

Biophilic Cities Maximize Moments of Awe

Meaningful engagement of nature includes watching and experiencing the many moments of wonder and awe that exist over the course of a day. Awe has been described by Louv as "what we feel during or after an encounter with something unexpected, and stimulates a sense of vastness and possibility, such as hearing thunder, listening to a moving piece of music, sensing the infinite during prayer or meditation" (Louv 2018). The experience of awe is now being actively studied with interesting results about its benefits. One recent study, for instance, demonstrates the power of awe to induce feelings of generosity and kindness, to help break out of our narrow self-perspectives, and to encourage pro-social behaviors (e.g. Piff et al. 2015).

There are already many abundant opportunities in cities to experience awe, from watching a diving peregrine falcon, to closely observing urban ants and other arthropods, to keenly watching clouds and weather. Biophilic cities work to expand these opportunities, and indeed one measure of a biophilic city is a city that works to maximize the potential moments of awe available over the course of the day.

In New York City, as one very important emerging example, residents now have the possibility of seeing a humpback whale, perhaps even (as some residents have enjoyed) from a window in one of the city's ferries. The number of whale sightings has increased dramatically in recent years, a function of improving the water quality of harbor and abundant menhaden fish stocks, a main food source for the whales. A nonprofit, Gotham Whale, has emerged as an advocate for the whales and runs a new citizen science initiative aimed at educating and engaging residents in this source of urban awe (see the new film about Gotham Green, Biophilic Cities n.d.-a, b, c).

Making Room for Other Forms of Life

As the story about New York City whales suggest, cities can and must be seen as places of habitat. Much of that magic is delivered in the form of biodiversity, and cities already harbor a great abundance of flora, fauna, and fungi. The aspiring biophilic city seeks to make room for even more. It is partly a recognition of the growing importance of cities in conserving biodiversity and partly the view that experiencing many other forms of life is uplifting and emotionally beneficial, often awe-inspiring. Members in the Biophilic Cities Network together account for a remarkable set of stories of coexistence and support for urban wildlife, from Austin's (Texas, USA) 1.5 million Mexican free-tailed bats, to Portland's (Oregon, USA) tradition of watching the evening spectacle (every night in September) at the Chapman School as thousands of Vaux's swifts roost in the school's chimney. Last year we had the chance to film this remarkable spectacle and the dramatic and sometimes quite loud human affirmation of the beauty and wonder of these birds.

In Singapore, they have witnessed the return of smooth-coated otters, which have become beloved, in large part the result of a major river restoration project in Bishan Park. Wolves in Oslo (Norway) have returned to the city's forests, and the city is developing a strategy to protect them and coexist with them. Several cities have now adopted bird-friendly design standards (e.g., San Francisco, Toronto), recognizing both the conservation imperative and the many important ways that birds enhance the quality of urban life. From mountain lions to wild boar to marine mammals, cities are increasingly working to celebrate, protect, and actively coexist with other forms of life.

Can cities be designed with wildlife and wild nature in mind? Edmonton, Canada, emphasizes ecological connectivity, extensively utilizing wildlife passages there. There is much more to do and even more radical ideas to consider, including the groundbreaking work of architect Joyce Hwang, who advocates for what she calls "habitecture" (e.g., design of facades and building surfaces to harbor a variety of life-forms; see Hwang 2017). Buildings like the Vancouver Convention Center have redefined their rooftop as a meadow and important bird habitat (with some 400,000 native plants; Kerr 2018). From swift boxes to hedgehog highways (now finding their way into major development projects in the UK), the design of cities and built environments will (and must) incorporate native flora and fauna.

We can rethink and reimagine cities as magical places and take important steps to protect, restore, and highlight the biodiversity with which we co-occupy the urban realm. London is being reimaged now as the world's first "National Park City," an idea of geographer Daniel Raven-Ellison. His epiphany was realizing that about half of greater London was taken up in greenery and nature; but not so much in conventional parks in backyards and other spaces between buildings. What it will mean to be a national park city remains unclear, but it has been changing local perceptions of the urban environment.

Reconceptualizing city as a national park can certainly serve to positively change the collective sense of the city. We need to do a better job at preparing the maps by which we signal and inform what is around us. Perhaps we need to redesign many of our urban maps so that they are better able to convey a sense of the mystery and wonder of the nature around us.

Biophilic Cities Is About Humans Caring About Nature

Biophilic cities are more than simply natureful; they are cities where humans actively engage in that nature in a variety of ways. Passive watching is one way, as when we see or hear birds over the regular course of the day. Ideally residents engage in that nature more actively, by exploring, by being curious and by learning the names of flora and fauna, and hopefully in the process developing a sense of caring about this nature. Even more active engagement can occur through citizen science programs and through volunteering in nature restoration and clean-up initiatives. Involvement by citizens in more nature clubs or nature activities is further evidence of biophilic engagement.

Cities are undertaking a variety of creative ways to engage residents in the nature around them, including BioBlitzes, creative biophilic maps and mapping, biophilic action pledges (e.g., Reston, Virginia's campaign to entice residents to commit to at least six biophilic actions; see Reston n.d.).

Nature Is a Birthright: Toward Just Biophilic Cities

Social equity has emerged importantly as a key concern when it comes to nature in cities. Thirty years ago Robert Bullard's book *Dumping in Dixie* (1990) helped to expose the unjust distribution of pollution and other environmental harms, showing that the chance of living near a hazardous waste dump or a superfund site was significantly higher for poor and minority communities. More recently there is a growing body of research that shows how unfair the distribution of environmental positives is. In most cities, access to nature is correlated with economic income. Similarly, tree cover is inequitably distributed, with poor neighborhoods having less tree cover than rich neighborhoods (Gerrish and Watkins 2018). Nature inequality extends beyond income, to racial and ethnic disparities. In many cities, minority or disempowered communities have less access to parks and less tree cover.

The inequality in tree cover has been perhaps most studied in the United States. For instance, a study of seven municipalities found that poorer neighborhoods have fewer trees than more affluent neighborhoods (Schwarz et al. 2015). Another study looked at ten municipalities, finding that neighborhoods with lower education and income had less vegetation, and that this trend that was largest in magnitude in large cities (Nesbitt et al. 2019). Similar trends have been observed using remotely sensed proxy variables for greenness (Casey et al. 2017). These differences in greenness matter because they are correlated with differences in surface temperature. For instance, a recent study of 108 US municipalities showed that historically minority areas have higher surface temperature (Hoffman et al. 2020), presumably in part because they have lower tree cover.

Nature inequality matters because where people live is determined by their income, and this in turn affects the provision of nature's benefits. Therefore, whatever mental health benefit comes from nature exposure, we may be providing to more affluent neighborhoods, so the rich live in cooler, less stressful neighborhoods. Conversely, poor neighborhoods do not have these benefits of nature, so they are disproportionately hotter and more stressful. When one asks about whether there is enough nature in cities for health, the current inequitable distribution of urban nature thus inevitably raises issue of race and class.

Many cities have embarked on programs to expand parks and greenspaces in neighborhoods of color and to help correct and rectify these inequalities. A fair distribution of nature is a key principle of biophilic cities. Nature is a birthright that every urban resident is entitled to enjoy. This is the key idea behind New York Nature Goals 2050, a grassroots initiative involving some 78 organizations that have together prepared an agenda for the preservation and enhancement of nature in that city. A key part of this effort has been a "Declaration of Rights to Nature," stating the essential notion that contact with nature is something every individual should enjoy. As co-founder Bram Gunther says, "nature is not ancillary, nature is not something you get because you lucked out… Nature is your right. Good nature is your right" (Gunther and Sanderson 2019). The NYC Nature Goals is at once a citizen planning process that has generated a set of 25 nature targets for the city, and also a process for activism, for citizens to fight for implementation of these targets. Citizens are asked to sign the online declaration of rights, with the goal of 10,000 signatures (Table 5.2).

Many cities have formerly expressed or stated nature equity goals in plans and ordinances. The city of Pittsburgh, for example, has stated the goal of ensuring every resident of the city is within a ten-minute walk of a quality park. Mostly this is a goal that has been achieved (over 80% of the city reaches this target), but even when a park is nearby it may be in disrepair. In recognition of the need to overcome years of disinvestment in the city's parks, and to address the park deficiency that exists for some

Table 5.2 Declaration of rights to New York City Nature

"New Yorkers need and have rights to a local environment that is healthy and whole, which provides safety, respite, and connection to the long history of life on Earth, rooted in the particular circumstances of its place. Such rights are essential to each individual as part of the community of nature as a whole. In spite of the biological richness of New York City at the time of its founding, the destruction of its natural resources in ways both large and small has been assumed to be an unfortunate but unavoidable cost of urban life. It is now self-evident that this assumption reflects a deep inconsistency with our ambitions, our ideals, and the facts: nature exists in the city, and given attention, management, and investment, can thrive again. We have an obligation to work together to create the social and economic circumstances necessary to lay the foundations for the long-term sustainability of the city's nature."

Source: NYC Nature Goals 2050 (n.d.)

neighborhoods, residents passed a ballot measure in November 2019, that will generate extensive new funding for parks.

Similar goals have been pursued in Portland. The recent opening of Cully Park, in a neighborhood of color, reflects the goal of providing adequate parks and greenspaces to all residents. Cully further reflects a commitment to engaging the neighborhood fully in the design of the park. Several creative engagement strategies were used, and the community was given an unusual degree of control over the design of the park. A large area of the park has been set aside as a community garden, for example, and these were directly designed by children from the neighborhood (Fig. 5.2).

Biophilic Cities Support Landscape Conservation

The vision of a biophilic city also includes care and concern for the more distant nature that exists beyond the boundaries of the city. If love of nature is the organizing idea of biophilic cities, then certainly that love must include global nature as well. Moreover, we know that in many ways the consumption patterns of urbanites have a large impact on global biodiversity. As a result, some cities have set goals and targets to reduce their ecological footprints. Setting ambitious local climate change targets would be another example. Pittsburgh, for example, has set the goal of 50% reduction in its greenhouse gas emissions by 2030 (Morrison 2019).

Fig. 5.2 The community garden in Cully Park, Portland, Oregon (USA). (Photo credit: Timothy Beatley)

Targets and Metrics

It is sometimes said that we measure what we care about. Adopting metrics and targets is an important step for cities. Progress toward the biophilic cities vision will be difficult to see without tracking these metrics.

There are a variety of examples of the targets and metrics that cities have adopted. Table 5.3 provides a comprehensive list and at least one way to organize or categorize different nature-based indicators (adapted from Beatley 2011). Cities today have commonly established targets that pertain to the extent of nature (e.g., the percentage tree canopy cover, parks, and greenspace acreage per capita) and access to nature for residents (e.g. percent of residents living within a ten-minute walk of a park or greenspace). Cities might also judge progress by the number of biophilic buildings or projects, or design features such as green walls or green roofs. For some cities it has been helpful to assess and track the number of residents actively involved (over the course of a year) in nature programs or activities.

Biophilic cities will judge their success in different ways, of course. Finding creative and useful ways to measure and track some of the

Table 5.3 Indicators of a biophilic city

How natureful is your city? Biophilic conditions and infrastructure
Percentage of population within a short distance of a park or greenspace. For example, Washington, DC, set a target of all residents within 10-minute walk of "parkland and Natural space" by 2032.
Existence of a connected, integrated ecological network; green urbanism from rooftop to region
Percentage of city land in wild or semi-wild nature
Percentage forest cover in the city. For example, Norfolk has a goal of reaching 30% tree canopy by 2040, up from its current 26%.
Extent and number of green urban features (e.g., green rooftops, green walls, trees)
Miles per capita of walking trails
Number of community gardens and garden plots (absolute and per capita)
Access to community garden area
How engaged and nature-active are residents of your city? Biophilic activities
Percentage of population that is active in nature or outdoor clubs or organizations
Number of such organizations active in the city
Percentage of population engaged in nature restoration and volunteer efforts as well as absolute number
Percentage of time residents spend outside (may vary depending on climate)
Percentage of residents who actively garden (including balcony, rooftop, and community gardens)
Extent of recess and outdoor playtime in schools
How knowledgeable and nature-literate are residents of your city? Biophilic attitudes and knowledge
Percentage of population that can recognize common species of native flora and fauna
Extent to which residents are curious about the natural world around them
What is commitment and capacity of your city to protect and restore nature. Biophilic Institutions and Governance
Adoption of a local biodiversity action plan or strategy
Extent of local biophilic support organizations, such as the existence of an active natural history museum or botanical garden
Priority given to environmental education
Percent of local budget devoted to nature conservation, recreation, education, and related activities
Adoption of green building and planning codes, grant programs, density bonuses, greenspace initiatives, and dark-sky lighting standards
Number of city-supported biophilic pilot projects or initiatives

Adapted and shortened from Beatley (2011)

essential elements of the biophilic vision—immersive nature, for example—will likely remain a challenge. Will one be able to experience nature on an hourly basis, as we hope? Will they be able to hear native birdsong on the walk to work or school (will you even be able to walk), and will you experience a sense of the awe and the vastness of the natural world in the course of a typical day? These questions may be more difficult to measure but are no less important.

Some Key Conclusions

This chapter has sought to paint the broader contours of what a biophilic city is and could be: what does it look and feel like, and what are its more specific qualities. We believe this is a compelling positive vision for future urbanization that allows for economically vibrant cities, full of interaction, that are also humane places to live. There are now plenty of good examples of how cities can be made more biophilic, and there is an ever-expanding toolbox of plans, codes, planning tools, and design ideas to apply. The next chapter discusses in more detail these more specific planning and design interventions.

References

Albro, Sandra. 2019. *Vacant to vibrant: Creating successful green infrastructure networks*. Washington, DC: Island Press.

Beatley, Timothy. 2011. *Biophilic cities: Integrating nature into urban design and planning*. Washington, DC: Island Press.

———. 2018. *Handbook of biophilic city design and planning*. Washington, DC: Island Press.

———. 2019. Designers walk: Toronto's new forest in the sky. *Biophilic Cities Journal* 3 (1): 24–25.

Biophilic Cities. n.d.-a. Connecting cities and nature. https://www.biophiliccities.org/.

———. n.d.-b. Perth urban wetland. A Film. https://www.biophiliccities.org/bcfilms.

———. n.d.-c. Gotham Whale. https://www.biophiliccities.org/bcfilms.

Bullard, Robert D. 1990. *Dumping in dixie: Race, class, and environmental quality*. Boulder, CO: Westview Press.

Casey, Joan A., Peter James, Lara Cushing, Bill M. Jesdale, and Rachel Morello-Frosch. 2017. Race, ethnicity, income concentration and 10-year change in urban greenness in the United States. *International Journal of Environmental Research and Public Health* 14 (12): 1546.

Children and Nature Network. n.d. Cities Connecting Children to Nature. https://www.childrenandnature.org/initiatives/cities/.

Connecting Nature. n.d. Bringing cities to life, bringing life into cities. https://connectingnature.eu/.

European Commission (EC). n.d. European Green Capital Award. https://ec.europa.eu/environment/europeangreencapital/.

Gerrish, Ed, and Shannon Lea Watkins. 2018. The relationship between urban forests and income: A meta-analysis. *Landscape and Urban Planning* 170: 293–308.

Goldsmith, Stephen. 2018. A data driven approach to cooling a city. Governing, December 18.

Gunther, Bram, and Eric Sanderson. 2019. Presentation at the nature of cities summit, Paris, June, 2019.

Hoffman, Jeremy S., Vivek Shandas, and Nicholas Pendleton. 2020. The effects of historical housing policies on resident exposure to intra-urban heat: A study of 108 US urban areas. *Climate* 8 (1): 12.

Hwang, Joyce. 2017. Toward an architecture for urban wildlife advocacy. *Biophilic Cities Journal* 1 (1): 24–31.

International Council for Local Environmental Initiatives (ICLEI). n.d. Cities with Nature. https://cwn.iclei.org/.

International Union for the Conservation of Nature (IUCN). n.d. Nature-based Solutions. https://www.iucn.org/commissions/commission-ecosystem-management/our-work/nature-based-solutions.

Kellert, Stephen, Judith Heerwagen, and Martin Mador, eds. 2011. *Biophilic design: The theory, science and practice of bringing buildings to life*. New York: Wiley.

Kerr, Jessica. 2018. Here's how the largest green roof in Canada gets its annual mowing. *Vancouver Courier*, October 18. https://www.vancouverisawesome.com/vancouver-news/green-roof-mowing-vancouver-convention-centre-1940610.

London National Park City. n.d.. http://www.nationalparkcity.london/.

Louv, Richard. 2018. 12 principles of a nature-rich city. https://www.childrenandnature.org/2016/10/18/12-principles-for-a-nature-rich-city/.

Morrison, Oliver. 2019. Pittsburgh's city government says it's on track to meet climate goals in public operations. But what about the rest of the city? *Public Source*, September 16. https://www.publicsource.org/pittsburghs-city-government-says-its-on-track-to-meet-climate-goals-in-public-operations-but-what-about-the-rest-of-the-city/.

Natural Capital Coalition. n.d. Cities. https://naturalcapitalcoalition.org/tag/cities/.

Nesbitt, Lorien, Michael J. Meitner, Cynthia Girling, Stephen R.J. Sheppard, and Yuhao Lu. 2019. Who has access to urban vegetation? A spatial analysis of distributional green equity in 10 US cities. *Landscape and Urban Planning* 181: 51–79.

Newman, Peter, and Jeff Kenworthy. 2015. *The end of automobile dependence*. Washington, DC: Island Press.

NYC Nature Goals 2050. n.d. NYC Nature Goals 2050. https://naturegoals.nyc/.

Piff, Paul, et al. 2015. Awe, the small self, and prosocial behavior. *Journal of Personality and Social Psychology* 108 (6): 883–899.

Portland Audubon. n.d. Backyard habitat certification program. https://audubonportland.org/get-involved/backyard-habitat-certification-program/.

Reston, Virginia. n.d. Reston biophilic pledge. http://www.reston.org/Portals/3/2020%20General/biophilic%20rack%20card%202020%20.pdf.

Schwarz, Kirsten, Michail Fragkias, Christopher G. Boone, Weiqi Zhou, Melissa McHale, J. Morgan Grove, Jarlath O'Neil-Dunne, Joseph P. McFadden, Geoffrey L. Buckley, and Dan Childers. 2015. Trees grow on money: Urban tree canopy cover and environmental justice. *PloS One* 10 (4): e0122051.

Singapore National Parks Board (Nparks). n.d. Park connector network. https://www.nparks.gov.sg/gardens-parks-and-nature/park-connector-network.

Stilgoe, John. 1998. *Outside lies magic: Regaining history and awareness in everyday places*. Walker Books.

The Nature Conservancy (TNC). n.d. Build Healthy Cities. https://www.nature.org/en-us/what-we-do/our-priorities/build-healthy-cities/.

The Nature of Cities (TNOC), collective blog. n.d. https://www.thenatureofcities.com/.

Ulrich, Roger. 1984. View through a window may influence recovery. *Science* 224 (4647): 224–225.

Urban Climate Lab. 2017. Dallas 2017 Urban Heat Island Management Study, along with Texas Trees Foundation. https://www.texastrees.org/wp-content/uploads/2019/07/Urban-Heat-Island-Study-Final_Print-Logos.pdf.

US Environmental Protection Agency (USEPA). n.d. Green Infrastructure. https://www.epa.gov/green-infrastructure.

Wilson, E.O. 1984. *Biophilia*. Cambridge, MA: Harvard University Press.

World Resources Institute. n.d. Cities4Forests. https://www.wri.org/our-work/project/cities4forests.

CHAPTER 6

Innovative Biophilic Design and Planning: From Rooftop to Neighborhood to City

Abstract Biophilic cities pursue the goal of abundant nature at multiple scales. This chapter examines in detail the best practices and design ideas that have been applied at the scales of building, neighborhood, and city-region. It is argued here that nature can be designed in at each scale, but ideally a biophilic city aspires to the interconnection of these scales. This chapter also examines the different plans that typically exist in cities and the places where biophilic elements might be found. The chapter also discusses the importance of local biophilic codes and provides examples of some of the most innovative of these.

Keywords Biophilic cities • Building codes • Green roofs • Trail networks • Urban forest plan • Zoning codes

This chapter focuses on showing inspirational, real-world examples of biophilic design and practice. Examples will be provided from multiple scales: building or site; neighborhood or eco-district; and city and regional/bioregional. The projects and examples of good practice reviewed here will provide the reader with a sense of the importance of integration across scales and the important ways in which the agenda of biophilic urbanism can be understood as "room or rooftop, to region or bioregion," and all the scales in between.

R. McDonald, T. Beatley, *Biophilic Cities for an Urban Century*,
https://doi.org/10.1007/978-3-030-51665-9_6

These will be compelling stories that will demonstrate what is possible and what cities around the world have been able to do to protect, restore, and celebrate the nature around them. This chapter will also review some of the different strategies to funding the implementation of biophilic urbanism. The chapter will also discuss some of the key challenges faced in biophilic cities, including political, legal, and institutional obstacles.

Building- and Site-Scale Biophilic Innovations

Biophilic design at the scale of buildings and sites has the potential to profoundly shape the environments in which we spend most of our time. There is a growing consensus that interior living and work environments can be much healthier, more stimulating, and psychologically restorative when efforts are made to design nature-in. Key design elements include extensive daylighting, natural ventilation, access to windows (preferably openable) with views of nature, as well as green walls, plants, and other forms of living nature. Other increasingly common design elements include biophilic materials such as wood and stone, and nature-inspired art.

A biophilic city is a city full of biophilic buildings. Every new building and every newly renovated existing building include to the extent possible biophilic design principles, seeking to connect its occupants and users to the natural world.

Biophilic buildings increasingly seek to overcome indoor-outdoor barriers, with green rooftops, skyparks, and green terraces that help bring nature inside the building and create important exterior green elements enjoyed by those outside the building. Creative examples include the new office project 300 Lafayette in the Soho district of New York City, where 11,000 square feet of planted green terraces are a key biophilic design quality and will bring nature inside but also create a significant biophilic experience for those on the outside looking in (e.g., see Beatley 2017).

Biophilic design elements such as green rooftops and living walls were once unusual but have now become mainstream in many cities. In Washington, DC, there are now more than four million square feet of installed green roofs. Green roofs continue to be created in Washington, as a function in part of city programs such as a financial rebate, stormwater management requirements, and a stormwater retention credit trading program.

Increasingly we are seeing urban rooftops become important places for food production, creating an opportunity for residents to be personally engaged in growing some of the food they eat. Via Verde, an affordable housing project in the Bronx, incorporates food production on one roof level through raised bed gardens and on another distinct level through the planting of an orchard.

The opportunities for green roofs are many, especially the rooftops of hospitals and medical centers. In Singapore, the biophilic Khoo Tech Puat Hospital or "KTPH" (Kellert Biophilic Design Award Winner) boasts an extensive rooftop farm, including 140 fruit trees (Beatley 2017). In America, the Boston Medical Center has installed a rooftop farm, producing some 5000–7000 pounds of fresh produce yearly, some going to inpatient meals, some going to the hospital cafeteria, some to a food pantry, and some to a demonstration kitchen. Instead of the 1500 miles or more that food typically travels from where it is produced to where it is consumed, in the case of BMC, it is but 1000 feet from the rooftop farm to the hospital's cafeteria. The hospital sees the rooftop farm as a positive feature providing multiple benefits. "The farm reduces the hospital's carbon footprint, increases green space, and reduces energy use, including the energy required to transport food" (Boston Medical n.d.).

Including biophilic features will significantly increase worker productivity and enhance quality of life. Increasingly, we recognize that biophilic design elements will significantly reduce the building's ecological and energy impacts. One good example is the Bullitt Center, a six-story office building in Seattle, which has served as the home for the Bullitt Foundation and the offices of the ILFI (International Living Futures Institute). The Bullitt Center was one of the first buildings to gain full certification under the Living Building Challenge (more rigorous than USGBC's LEED program, and with an explicit focus on the biophilic qualities of buildings; see Sturgeon 2017).

The Bullitt design is groundbreaking in many ways. Its visually dramatic overhanging roof hosts photovoltaic panels, 575 in all. These, together with the building's low-energy design (including extensive natural light and openable windows), produce more energy than the building needs over the course of the year. The building collects, treats, and reuses all of the rainwater that falls on the site. In this way it functions as a stand of native Douglas Fir would have prior to settlement. One level of the building is a wetland that treats all the gray water from the building. Black water is treated through use of waterless composting toilets that makes agricultural compost from this waste stream.

Wood dominates the natural materials used in the Bullitt Center, delivering impressive biophilic vistas for workers and visitors. One is surrounded by the warmth of this wood on ceilings and floors. The wood is FSC-certified, sourced from within a 500 mile radius of the site. The building uses no materials from the so-called Red List ("worst in class materials prevalent in the building industry," ILFI n.d.). The Red List includes such substances as chlorinated polymers such as PVCs, chemicals like formaldehyde, and toxic heavy metals like arsenic, cadmium, lead, and mercury.

The Bullitt Center illustrates the important ways in which biophilic design and design for health overlap. One prominent feature of the building is what is sometimes called the "irresistible staircase." The building has elevators, but occupants and visitors are encouraged to use the more central and visually prominent stairs—with beautiful wood and window views of the Seattle skyline, the goal was to entice the use of the stairs (Fig. 6.1). Wood is widely viewed as a biophilic material, and it is worth noting that its use is rising in the sustainable design world. There are now several larger, high-rise structures that have been built utilizing the technology of cross-laminated timber, which is increasingly popular. At least theoretically, wood is a sustainable resource, and in tall timber structures offers the possibility of significant sequestration of carbon, as well as the avoidance of high-energy and high-carbon materials such as steel and concrete.

Retrofitting existing structures in the city offers a similar benefit in preserving embodied energy. As historic preservationists often say, the most sustainable building is an existing building. Biophilic and natural design elements can be integrated into structures that might not have been very biophilic in their original form. The new Atlanta headquarters of Interface Carpets shows precisely the power of retrofitting structures in ways that include biophilic features. Affectionately referred to as the "Basecamp," the building models itself after the form and function of a forest—in this case an eastern US deciduous forest. Its most dramatic biophilic feature is the building's outer facade, essentially a large image of an eastern US forest, consisting of a printed fritted surface that allows workers to see out and the public to see in. It is a visually striking element that has already made the building a prominent landmark in midtown Atlanta.

Pittsburgh is the home of several recent excellent examples of biophilic design, including the Phipps Conservatory's Center for Sustainable Landscape (CSL) and the new Frick Environmental Education Center. Each is a certified Living Building and includes impressive biophilic

Fig. 6.1 The Bullitt Center's living staircase, in Seattle, Washington (USA). (Photo credit: Timothy Beatley)

qualities. In the CSL, employees sit at desks that are never far from operable windows, with spectacular views of surrounding nature, their interior workspaces drenched in daylight. There is an emphasis on biophilic inspiration, and the wonderful green roof is open to the public. The CSL project emphasizes the use of native plants throughout, and like other living buildings is net zero for energy and water.

Many of the best examples of biophilic design seek to bring nature inside, often in very creative ways. The Frick Center carries a wood and forest theme throughout its interior. The windows and facade paneling were designed intentionally to mimic the diversity of a forest. The translucent PV panels that generate all the power needed for the site are meant to mimic the dappled light of the forest (for more about Frick, see Sturgeon 2017).

Biophilic Design at the Neighborhood Level

While we spend much of our life in buildings, these buildings—homes, offices, schools—are embedded in larger neighborhoods that tend to physically and socially structure our lives. These assemblages of buildings, plazas, shops, and streets often have special meaning and coherence to us. Neighborhoods are just the right scale for meaningful biophilic interventions, whether expressed in the planting of street trees, the creation of new pocket parks, or the design of a new library. It is often at the urban district or neighborhood scale that some of the most impressive and impactful actions can be taken to make nature a more central aspect of the lives of urban residents.

It is especially important to understand what can be done to support nature beyond buildings, to think creatively about the many spaces around and between structures. Here there will be many opportunities to insert nature into these places. Sometimes an opportunity exists to design and build an entirely new biophilic neighborhood, but more often it is the chance to retrofit an existing urban neighborhood in ways that make it more natureful.

In many older American cities, especially the legacy cities of the Midwest—cities like Cleveland, Youngstown, and Detroit—there are immense opportunities to repurpose vacant land for nature. BC Partner city Milwaukee has been implementing for several years an innovative program called HOME GR/OWN that aggregates vacant lots and repurposes them as pocket parks and orchards (See Beatley 2017). Recently the city has designed these parks to serve multiple community functions, including important stormwater resilience functions. One impressive recent example is Fondy Park, adjacent to Fondy market, which has been redesigned as a wildflower meadow that can collect and retain more than 70,000 gallons, contributing to the city's larger effort at managing stormwater.

In Rotterdam (Netherlands), a city that has emerged as a laboratory for new ideas about resilient urban planning, new neighborhood squares are designed as important rainwater retention facilities, part of the city's larger effort to address resilience (and in support of its goal of becoming a Climate Proof city by 2025). The so-called Water Plazas have been designed to add new community spaces, but also to collect and retain water during rain events. These spaces include extensive plantings as well as new community sports and gathering spaces, all designed to collect and retain rainwater. The Dutch and Danish have together pioneered the idea of community spaces that also act as flood and rainwater retention facilities.

Cities sometimes have the chance to create entire new neighborhoods when larger blocks of land become available. Austin, Texas (USA), for example, has redeveloped its former Mueller Airport, an area of more than 700 acres in the city, into a new neighborhood, exemplifying the design qualities of "green urbanism." The urban form reflects the principles of new urbanism, with connected streets and sidewalks, as well as extensive parks and greenspaces. While not yet fully built-out it will include nearly as many jobs as housing units, all embedded in a larger neighborhood context of trees and nature.

The legal and planning status of neighborhoods varies from city to city. In some cities, such as Seattle, Washington (USA), there is an extensive history of planning efforts at the neighborhood level, and even an Office of Neighborhood Planning to help. In Seattle, neighborhoods are empowered to directly develop their own visions of the future, deciding themselves through extensive community participation how they wish to accommodate expected future growth, which is assigned to them by the city's higher-level comprehensive plan.

Biophilic urban design at the neighborhood or district scale offers many opportunities. While achieving a fully nature-immersive city will be difficult and likely not achievable in the short term, achievement of such outcomes in a single urban neighborhoods is more realistic. It is also at the neighborhood level where the impact of biophilic design is most easily felt, where the practical possibilities more easily appreciated, and where residents can be directly engaged. Many biophilic design ideas are possible here, large and small: daylighting streams and installing green streets, repurposing vacant spaces, planting gardens of every type, creating pollinator pathways and new habitats for biodiversity. Neighborhoods can be designed in ways that facilitate engagement with the nature around—including nature discovery stations, neighborhood libraries that are equipped with telescopes and microscopes, and curated self-guided nature hikes.

It is useful to think of the urban neighborhood as a spatial frame that may help to tie together and amplify the impact of a rising number of biophilic buildings. Multiple green rooftops, for instance, are an opportunity at a neighborhood scale to create the stepping-stone habitats for birds and other urban wildlife. Equally true, the surrounding biophilic qualities of a neighborhood can help to reinforce the nature-connection goals of a biophilic building. As we have seen, biophilic design emphasizes natural light and extensive windows. View of nature from windows is an essential biophilic quality, requiring actions at the neighborhood level to enhance the natural environment beyond buildings.

In Portland, Oregon (USA), they have reshaped certain districts around biophilic and ecological principles. For instance, the Lloyd District, a commercial area of the city, containing among other things the Oregon Convention Center, has been the focus of innovative planning efforts over the years. It has been designated an "EcoDistrict," one of the first in the country. An innovative plan (called "roadmap") was developed for the district in 2012, organized around four "big ideas," one of which was the idea of Lloyd as a "Biophilic District." The plan identifies specific elements of the biophilic neighborhood and goals for the future, including new green streets and parks, as well as ambitious limits on impervious surfaces (Portland Sustainability Institute 2012). Smaller pilot projects have occurred in the Lloyd District, including an innovative wastewater reuse plan, *Hassalo on 8th*, a system covering a four-block area designed by Biohabitats. Wastewater from three buildings (45,000 gallons a day) is treated through "trickling filters and constructed wetlands," resulting not only in the diversion of wastewater but also in the introduction of a significant biophilic green element into the neighborhood (Biohabitats n.d.). And in many Portland neighborhoods "green streets" have been installed that help collect and retain stormwater and to add an element of greenery (Fig. 6.2).

Some of the most impressive biophilic neighborhoods have been urban districts in European cities. Many of these have been extensively studied and written about, including Vauban, a wonderful car-free neighborhood in Freiburg, Germany, and the Western Harbor, a pioneering green neighborhood in Malmo, Sweden. Both appear in the documentary film "The Nature of Cities." At Vauban, the neighborhood design includes extensive areas of public greenspace and play areas in the form of courtyards that are off-limits to cars, along with efforts to protect existing trees. This relatively dense new neighborhood is served by a new tram line and easily

Fig. 6.2 An example of green stormwater infrastructure in Portland, Oregon (USA). (Photo credit: Timothy Beatley)

reachable by bicycle (Freiburg as a city has a remarkable history of prioritizing for bikes and walking; See Medearis and Daseking 2012).

Another recent example of a biophilic neighborhood design can be seen in the community of Serenbe, outside Atlanta, Georgia (USA). Perhaps more the model of a village than an urban neighborhood, Serenbe provides residents there with an almost unequaled access to nearby nature.

A compact and highly walkable community (with 70% of the land area set aside for conservation), "omega-shaped" neighborhoods ensure that residents have one side facing the sidewalks and built environment of the village and the other side facing trees, trails, and farmland (Tabb 2016).

Many of the members of the Biophilic Cities Network have adopted creative approaches for greening neighborhood spaces, taking up paved hard surfaces and replacing them with plants and trees. Few cities have done as much as San Francisco, which provides opportunities for residents to install sidewalk gardens thorough a special Sidewalk Landscaping Permit. More than 2000 of these have been issued. Living Alleys are another innovation in San Francisco and are making their way into the city's area plans. Another innovation is the city's unique Street Parks program, which allows neighborhoods to convert median strips into community gardens and gathering spaces (there are now more than 100 designated street parks; e.g., see San Francisco Public Works n.d.).

One lesson is that cities can often create opportunities for new greenspaces, while efforts are made to moderate the outsized impact of cars and car traffic. Partner city Barcelona's experience with superblocks provides another inspiring and creative approach to bringing more nature into an urban neighborhood. The brainchild of Salvador Rueda, the city longstanding director of Urban Ecology Agency, the superblock idea is essentially to route higher-speed car traffic around superblocks, significantly calming through-traffic in the interior neighborhood living spaces of these superblocks (Roberts 2019). The results in the first six superblocks are impressive indeed. They allow for the creation of new public spaces and new places for planting vegetation. A recent study applying health impact assessment methods concludes that if the city completes its planned 503 superblocks, the health benefits will be considerable: an estimated 24% reduction in air pollution and the prevention of some 667 premature deaths each year. Superblocks have also been shown to reduce noise and to increase physical activity of residents (Mueller et al. 2019).

Cities like Oslo (Norway) have taken similar steps to reduce the impact of automobiles and to create car-free zones, now including much of its downtown. In addition to the opportunities to provide more space for nature, such efforts can also make cities much safer. Oslo has successfully implemented a comprehensive Vision Zero strategy for expanding bicycle lanes, pedestrian space, and public transit options. It has also reduced parking for cars and raised car tolls to discourage car use. In 2019, remarkably, there were no pedestrian or bicycle fatalities in this city of 670,000.

The Oslo and Barcelona experiences show compellingly that it is possible to create the conditions for residents to spend more time outdoors, in physically active modes of mobility, resulting in significant improvements in health and happiness.

Citywide Biophilic Planning Innovations

Every city consists of buildings and neighborhoods, of course. But biophilic cities are more than the sum of these parts. It is essential to take steps at planning and designing that work to advance a citywide, integrative vision. Moreover, many of the specific steps a city can take to strengthen its biophilic *bonafides* are policies, initiatives, and code requirements that apply to the entire city. In this section we discuss design and planning at this larger city scale.

Citywide strategies can seek to protect, restore, or celebrate ecological processes that bridge neighborhoods and help to weave the city together, environmentally and socially. A focus on hydrology is one major opportunity, and there are many cities that have made investments in their river and stream systems. Cities like Richmond, Virginia (USA), or Milwaukee, Wisconsin (USA), have done much to connect residents to their rivers, and hence to nature.

The presence of the James River, for instance, is a key defining citywide ecological feature for Richmond, VA (USA) (Fig. 6.3). As a result there are many opportunities to access wild spaces along the river, which has class IV rapids. A new riverfront master plan establishes the James River as Richmond's "Central Park" and lays out a series of projects to further expand these connections to water. Part of this master plan is to create abundant opportunities for residents to gain new physical access to this remarkable river. One of the newest biophilic projects has been the Tyler Potterfield Memorial Bridge, a pedestrian and bicycle bridge that crosses the river. The bridge is built on top of an old hydroelectric dam, and provides "unequaled river vistas, panoramic views of the downtown skyline, along with sounds that belie the extraordinarily close relationship of the bridge to the roaring river" (Friends of the James River Park n.d.).

Some cities such as Atlanta have invested in citywide park projects that serve to physically connect and tie together neighborhoods. Atlanta's innovative BeltLine, 33 miles of multiuse trails following old rail lines that encircle the city, is one such example, providing connection between 45 neighborhoods. The BeltLine serves to connect people

Fig. 6.3 A footbridge over the James River, in Richmond, Virginia (USA). (Photo credit: Timothy Beatley)

and parks and is itself a linear forest (the BeltLine Arboretum), as well as a display area for public art. Parts of the BeltLine will eventually serve as a transit corridor and a site for new affordable housing.

How Do Biophilic Cities Work Nature into Their Plans?

As these citywide initiatives suggest, biophilic principles, goals, and targets can make their way into a number of different planning documents. The panoply of plans and plan-like documents in a city can be complex and confusing. How cities plan often varies significantly from city to city, depending upon the planning system, laws, and traditions that exist in a particular jurisdiction. In the United States, most cities have developed comprehensive plans (also called master plans or general plans, as they are California) that are intended to guide spatial development and growth in the future, often with a timeframe of 30 to 40 years. Other substantively focused plans and strategies, such as a climate change plan or urban forest strategy, are often subsumed within a comprehensive plan.

Table 6.1 presents a rough typology of different types of city plans that have been employed, and some specific examples for each category. Some plans are focused on important ecological features in a city, such as rivers,

Table 6.1 A typology of local plans with examples from network cities

Comprehensive or general plans
Richmond 300
Functional or topical plans and strategies
Arlington county public spaces master plan
Vancouver bird strategy
Norfolk vision 2100 resilience strategy
RVAGreen 2050
Austin urban forest plan
Sustainable DC plan
Resource-specific plans and strategies
Toronto Ravine strategy
Richmond riverfront plan
Citywide action plans
Vancouver greenest city action plan

Source: Adapted and expanded from Beatley (2019)

canyons, and mountains. Toronto, for instance, is sometimes called the Ravine City, a reference to the important role this geologic and topographic element plays in defining the city. The city's ravines comprise remarkable 17% of the city's land area and represent a major unifying natural or biophilic feature, including the beloved High Park and other important spaces of nature. The Ravine Strategy is an example of one of several resource-specific plans recently adopted by the city. It lays out a vision for the Ravines, five guiding principles and twenty actions for the future (City of Toronto 2020).

Arlington County provides an excellent example of how ideally this panoply of plans can be hierarchical and interlocking. Its newly adopted Public Spaces Master Plan serves as a chapter in its comprehensive plan. Several more focused plans, such as Arlington's Urban Forest Master Plan, fit in as elements within and under the Public Spaces Master Plan. For Arlington the Public Spaces Master Plan is the place to lay out its vision for parks, greenspaces, and other civic space. The Plan identifies the goal of adding 30 additional acres of public greenspace over the next ten years, no small feat in a dense locality (Arlington County 2019).

One of the most interesting and innovative features of Arlington's Plan is its discussion of "casual use spaces," or "spaces that are intentionally designed to support casual, impromptu uses and connections with nature." Though these are not mapped, their characteristics and qualities are described (e.g., places for bird-watching and picnicking and strolling through the woods). The plan also contains targets for the community's off-street trail system (already over 50 miles in length), including adding important connections (creating an inner and outer trail loop). Not all goals are specifically about physical space, as the Plan also identifies the important goal of expanding the number of nonprofit and advocacy groups that work on (and raise funds to support) Arlington's public spaces.

Similarly, many cities have developed urban forest plans or strategies. Austin, Texas (USA), for example, adopted such a plan in 2013, laying out a 20-year vision and specific actions for managing trees on publicly owned land in the city. An important element of this urban forest plan is a report on the state of the forest. Often such plans estimate numbers of trees (33 million in Austin), their geographic distribution, tree canopy cover (38%), as well as goals and actions for the future (City of Austin n.d.).

Some of the most impressive elements of citywide biophilic vision can be seen in coastal cities like New York. New York City has continued to work on reconnecting the city to its water, especially through its groundbreaking vision *2020: Comprehensive Waterfront Plan*, adopted in 2011 (City of New York 2011). The results have been impressive and have helped the city to rightly regain its historic moniker of City of Water. New waterfront parks, such as Brooklyn Bridge Park and Hudson River Waterfront Park, aim to expand physical access to water. New efforts at making shoreline parks and spaces have, especially post Hurricane Sandy, emphasized their resilience and "floodability" (Fazzare 2018). Physical access to the water's edge, and ability to launch a kayak or canoe, is balanced against the need to allow and encourage periodic flooding.

This reconnection with the city's watery realm (what the City's waterfront plan refers to as the city's Sixth Borough) is further aided by the work of local partners such as the Waterfront Alliance, which has created their own voluntary shoreline development certification system (Called *WEDG: Waterfront Edge Design Guidelines*; Waterfront Alliance n.d.). This coalition of more than 1100 organizations has also been a fierce advocate for expanding the city's water-based transit system (including new ferry routes and service).

Few cities have been as innovative in marrying nature and resilience as Norfolk, Virginia. Its Vision 2100 plan is the key document. It divides the city into four vision areas depending upon assets and risks. The plan focuses infrastructure and other investments in the most resilient parts of the city, with new urban centers on the highest ground. An innovative resilience-based zoning code has been adopted to implement the Vision 2100 plan, and now requires all new development to meet a minimum resilience quotient, essentially a point system that gives credit for inclusion of risk-reducing biophilic design elements such as green roofs. The Norfolk example shows how a vision and practice of biophilic cities can help to make a city more resilient. It also shows the importance of zoning and other codes that enforce and implement the vision on the ground.

Biophilic Implementation Tools: Codes, Programs, and Initiatives

An important piece of any city's approach to creating and maintaining nature is the full package of ordinances and codes. Cities adopt a variety of laws and legal requirements, and often it is through these that biophilic requirements or incentives are implemented and enforced. Cities have adopted a range of nature-related ordinances, and these include standards and requirements for stormwater management, for protection of trees and other natural features, for bird-safe windows and facades, and for restrictions that pertain to livestock and urban agriculture. Sometimes these laws are integrated and aggregated into a larger municipal green code. The City of Toronto, for instance, bundles several biophilic standards and requirements, including a green roof mandate, tree protection standards, and dark sky lighting requirements.

Table 6.2 summarizes biophilic codes and initiatives in several key cities, with links to actual code language. Some of these—for instance—tree protection codes—are not especially new (though in many cities they remain controversial). Other biophilic standards are much newer, such as the growing trend of sitting requirements for bird-safe windows and building design.

A few cities are now considering adoption of bird-friendly design standards, which is a positive sign. It is estimated that as many as a billion birds die each year from building and window strikes. Birds have difficulty perceiving windows as barriers, leading to fatal collisions. The urban design answer is clear: there are now a variety of bird-window treatments, including fritted glass, that allow birds to perceive windows as barriers. San Francisco was the first American city to mandate bird-friendly building facades, but New York became the largest and most recent in December 2019, when its City Council adopted bill 1482B with a resounding vote of 43-3. The law, which takes effect at the end of 2020, requires the use of bird-friendly materials for the first 75 vertical feet of a building (where most bird strikes occur), as well as facades above green roofs.

While there was some resistance from the building sector, the point has been made that fritted and other bird-friendly glass is also more energy efficient, helping the city to move toward its ambitious carbon-reduction goals while creating a more bird-friendly city. The effectiveness of this approach can be seen in the retrofit of the Jacob Javits Center in New York. During a billion-dollar renovation, the decision was made to replace all

Table 6.2 Examples of biophilic city codes, policies, programs

Singapore	Landscape replacement policy	Regulations
Singapore requires most proposed large-scale buildings to replace at a minimum the nature lost as a result of the ground-level footprint of the structure, as a part of its LUSH initiative (Landscaping for Urban Spaces and High Rises). Vertical greenery is provided in the form of skyparks, green rooftops, and other green features, and is counted by measuring the extent of their horizontal area. In addition, buildings must also meet a minimum Green Plot Ratio requirement (which a ratio of total leaf area to site area, described as a measure of the "density of greenery). These requirements have led to a friendly competition among designers and developers to see which buildings can boast the most vertical nature and highest ratio of vertical/replaced to lost/ground-level nature. More information: Singapore Nparks (n.d.); see also URA (n.d.).		
Washington, DC	Stormwater retention credits	Financial incentive
An incentive program run by the Washington Department of Energy and Environment. Participants are incentivized to undertake urban greening projects—installing rain gardens, taking up impervious pavement, installing green rooftop—in exchange for stormwater credits which can then be sold to other individuals and companies to be used to satisfy or partially satisfy their stormwater management requirements. More information: City of Washington, DC (n.d.).		
Portland, OR	Green streets program	Program
A program initiated in 2003 as a pilot program which has now become a permanent initiative of the City of Portland's Bureau of Environmental Services (BES). The program supports the installation of "green streets," which are essentially portions of roadways and sidewalks converted to vegetative stormwater collection bioswales. There are now approximately 2000 green streets in Portland, financially supported in part through the city's "One Percent for Green" fund. More information: City of Portland (n.d.).		
Seattle, WA	Seattle green factor	Regulations
This regulation applies to new construction of commercial and multifamily residential. Utilizing a spreadsheet and point system, development must incorporate sufficient greenery and green design elements to achieve a minimum point total. More information: City of Seattle (n.d.).		
San Francisco, CA	Standards for bird-safe buildings	Regulations
San Francisco was the first American city to adopt minimum bird-safe design standards. The code requires the use of fritted glass and other bird-safe glazing materials for new construction or major renovations in bird collision zones (90% of glazing up to 60 feet) and for hazardous building features such as a glass skyway (100% treatment). Bird collision zones are areas adjacent to parks, greenspace, open water, and green roofs. Required window patterns must be 4 inches (tall) by 2 inches (wide), or smaller. The standards also address lighting and small wind turbines. More information: City of San Francisco (n.d.).		
Toronto, Canada	Toronto green standard	Regulations

(*continued*)

Table 6.2 (continued)

A comprehensive set of sustainable design standards. Tier 1 standards are mandatory, and include bird collision deterrence, tree planting, dark sky lighting, and minimum green roof installation requirements (a minimum of 50% of a new rooftop must be planted as "biodiverse green roof" in part to support pollinator species). More information: City of Toronto (n.d.).		
Austin, TX	Tree ordinance	Regulations
Austin has one of the oldest municipal tree protection codes in the country. Permits are required to cut trees over a certain diameter (19 inches) and subject to review by the city's forester. Special provisions apply in construction sites to protect root zones, and there are special (more stringent) provisions to protect Heritage Trees (trees over 24 inches in size, and on a list of specific species). Under Austin's tree protection regulations, Heritage Trees can be removed only when determined to be "an imminent hazard to life or property." More information: City of Austin (n.d.).		
Raleigh, NC	Capital area greenway	Program
The Capital Area Greenway System (CAGS) is described as a "network of linear parks, located along rivers, creeks and streams, that provide opportunities for conservation, recreation and connecting people to Raleigh's parks and other community features." The system now encompasses 28 trails covering 117 miles (paved and unpaved) and protecting some 3700 acres. More information: City of Raleigh, 2020.		

glass with fritted bird-friendly glass, as well as install a nearly seven-acre green rooftop. The results have been impressive: a New York City Audubon study reports a 90% reduction in bird fatalities as a result of the new glass. The Javits Center went from a bird death-trap to a success story, with birds nesting and fledging on the rooftop.

These examples show the power of these kinds of biophilic codes for helping to bring about natureful buildings, neighborhoods, and cityscapes. Cities have adopted a variety of innovative codes, from bird-friendly design, to mandates for green roofs and minimum greenspaces, mandates for stormwater management, and tree protection standards.

Connecting Cities to Continental-Scale Conservation Visions

It is important to recognize that cities are limited in their scope, subject to the boundaries of their specific jurisdiction. A city will have the opportunity to require the installation of green roofs within its territory but will have little ability to control what happens beyond its borders, even in neighboring jurisdictions. There are nonetheless some institutional structures available to overcome this form of biophilic localism. These include regional councils of government and, in some cases, regional governmental bodies (such as Portland's Metropolitan Services District) that have significant land use powers.

Many metropolitan areas have some form of regional or bioregional planning structure. The urbanized metro area of Toronto (Canada), for instance, is surrounded by the large Ontario Greenbelt, a layer of management and planning and protection for farmlands, wetlands, and important natural areas such as the Niagara Escarpment. Another example is Portland, Oregon (USA), where a Metropolitan Greenspaces Master Plan guides acquisition and management of a "regional system of natural areas, trails and greenways for wildlife and people in the Portland region" (Portland Metro n.d.).

Short of more formal engagement at the regional level, a city can work to educate other localities in the region in the vision and practices of biophilic cities, and the importance of nature-based solutions to regional problems. This might happen through regional forums, establishing procedures for consultation and sharing of experiences.

It is important also to recognize that cities can and should be thinking about the nature beyond their boundaries, in part because the actions and consumption patterns of residents have serious impacts on sometimes faraway ecosystems. The ecological footprint of urban residents in often quite large, and cities should consider policies that might reduce impacts, as well as support (politically and perhaps financially) conservation efforts beyond their borders.

Biophilic cities should work to understand how they fit into (and could better support) the larger ecological systems and landscapes in which they are embedded. The Wildlands Network has produced a remarkable and comprehensive vision for the conservation of land in the eastern US. They call it the Eastern Wildway, constructed largely by overlaying existing conservation land and imagining new ecological connections and buffers. This unusual map is "meant to inspire conservationists, it's meant to inspire citizens, [that] we could pull this off" (Sutherland 2019). Sutherland recognizes though that cities were not really a focus of this mapping effort. They are a missing but essential piece in this subcontinental land use and conservation vision. More can be done to ensure that the policies and land use of cities complement, and not work against, this vision.

Some Key Conclusions

We have attempted in this chapter to describe some of the main ways in which cities can protect, restore, grow, and celebrate the nature around them. There are now many cities that offer hopeful examples of what is possible. We have tried to describe some of these here, but there are many more around the world. Our discussion provides but a sampling of the work going on, and undoubtedly the number of urban exemplars and examples of creative projects, initiatives, and ordinances will increase over time as the vision of biophilic cities continues to gain traction.

Cities today and in the future will face a series of daunting challenges—from adaptation to climate change, to overcoming growing sedentary lifestyles and work patterns, to creating the conditions for mental health and well-being, to tackling a variety of other serious (local and global) environmental problems including loss of biodiversity and declining air quality. We argue that there are few solutions that will effectively (and simultaneously) address these myriad challenges in the way that nature can. Access to abundant nature in the growing cities of the world, more than almost anything else, can also deliver the joy and meaning we need in our lives.

References

Arlington County, VA. 2019. *Public spaces master plan*. https://projects.arlingtonva.us/plans-studies/comprehensive-plan/public-spaces-master-plan/.

Beatley, Timothy. 2017. *Handbook of biophilic city planning and design*. Washington, DC: Island Press.

———. 2019. Designers walk: Toronto's forest in the sky. *Biophilic Cities Journal* 3 (1): 22–24.

Biohabitats. n.d. Hassalo on 8th wastewater treatment & use. https://www.biohabitats.com/project/hassalo-on-8th-wastewater-treatment-reuse-system-2/.

Boston Medical. n.d. Rooftop farm. https://www.bmc.org/nourishing-our-community/rooftop-farm.

City of Austin. n.d. Tree regulations. https://www.austintexas.gov/faq/tree-regulations.

City of New York. 2011. *Vision 2020: Comprehensive waterfront plan*, March. https://www1.nyc.gov/assets/planning/download/pdf/plans-studies/vision-2020-cwp/vision2020/vision2020_nyc_cwp.pdf.

City of Portland, OR. n.d. Portland green street program. https://cityparksalliance.org/resource/portland-green-street-program/.

City of San Francisco. n.d. Standards for bird-safe buildings. https://sfplanning.org/standards-bird-safe-buildings.

City of Seattle. n.d. Seattle green factor. https://www.seattle.gov/sdci/codes/codes-we-enforce-(a-z)/seattle-green-factor.

City of Toronto. 2020. Ravine strategy. https://www.toronto.ca/city-government/accountability-operations-customer-service/long-term-vision-plans-and-strategies/ravine-strategy/.

———. n.d. Toronto green standard. https://www.toronto.ca/city-government/planning-development/official-plan-guidelines/toronto-green-standard.

City of Washington, DC. n.d. Stormwater retention credits. https://doee.dc.gov/src.

Fazzare, Elize. 2018. Hunters point South Park is a model for urban flood resiliency. *Architectural Digest*, June 27.

Friends of the James River Park. n.d. About the James River Park system. https://jamesriverpark.org/.

International Living Futures Institute (ILFI). n.d. The red list. https://living-future.org/declare/declare-about/red-list/.

Medearis, Dale, and Wulf Daseking. 2012. Freiburg, Germany: Germany's Eco-Capital. In *Green cities of Europe: Global lessons on green urbanism*, ed. Timothy Beatley. Washington, DC: Island Press.

Mueller, Natalie, et al. 2019. Changing the urban design of cities for health: The superblock model. *Environment International*. https://doi.org/10.1016/j.envint.2019.105132.

Portland Metro. n.d. Metropolitan greenspaces master plan. https://www.oregonmetro.gov/metropolitan-greenspaces-master-plan.

Portland Sustainability Institute. 2012. *Lloyd Ecodistrict Roadmap*. http://www.ecolloyd.org/wordpress/wp-content/uploads/2011/10/lloyd_roadmap_FINAL_hires.pdf.

Roberts, David. 2019. Barcelona wants to build 500 superblocks. Here's what it learned from the first ones. *Vox*. https://www.vox.com/energy-and-environment/2019/4/9/18273894/barcelona-urban-planning-superblocks-poblenou.

San Francisco Public Works. n.d. Street parks program. https://sfpublicworks.org/streetparks.

Singapore National Parks Boards (Nparks). n.d. Guidelines for landscape replacement areas within new developments in (part) downtown core, (part) straits view, (part) Kallang and (part) Jurong East planning areas. https://www.ura.gov.sg/Corporate/Guidelines/Circulars/dc09-09.

Singapore Urban Redevelopment Authority (URA). n.d. Development control. https://www.ura.gov.sg/corporate/guidelines/Development-Control.

Sturgeon, Amanda. 2017. *Creating biophilic buildings*. Ecotone Publishing.

Sutherland, Ron. 2019. Personal interview by Timothy Beatley, April.
Tabb, Phillip. 2016. *Serene urbanism: A biophilic theory and practice of sustainable placemaking*. Routledge.
Waterfront Alliance. n.d. *Waterfront edge design guidelines*. http://wedg.waterfrontalliance.org/.

CHAPTER 7

The Choice

Abstract This chapter presents a case study of the Los Angeles River, from its ecological death through its partial rebirth. We live in a world filled with many places like the LA River. If we continue building cities like this, we risk condemning billions of people to suffer from the urban mental health penalty. But cities have a choice of taking another, more biophilic path. This chapter discusses in detail some challenges to the biophilic cities vision that must be addressed if the vision is to move forward in cities around the globe. We end by listing what readers of this book can do to help move their city toward a more biophilic future.

Keywords Eco-gentrification • Global South • Groundwater • Los Angeles River • River restoration • Stormwater

The LA River Story

The Los Angeles River was once famously described by journalist Dick Roraback as narrowing "to a stream, then a creek, finally a joke" (Roraback 1985). For much of its length, the river has been altered by man into a concrete-lined channel, designed to carry storm and flood waters as quickly out to sea as possible. In movies and popular culture, the LA River has become a place for noir mysteries and car chases, a symbol of the dark side of a city. To environmentalists, it has become emblematic of a river that has lost all ecological and aesthetic function. Indeed, for several

R. McDonald, T. Beatley, *Biophilic Cities for an Urban Century*,
https://doi.org/10.1007/978-3-030-51665-9_7

decades, the river itself, as a real entity, seemed to disappear from the government's memory. The 1965 Master Plan of the city of Los Angeles doesn't even refer to a river, just an "open flood control channel" (Gandy 2014).

It wasn't always this way. Traditionally, the LA River was a perennially flowing stream in a semi-arid landscape. Fed by a modest amount of groundwater, it was a small stream lined with cottonwood and willows, albeit a stream that was prone to massive floods every decade or so when there was an intense rain event in its headwaters. Native Americans and then the first Spanish settlements occurred near this perennial water source, although usually just outside the floodplain. However, rapid urban growth in the late nineteenth and early twentieth centuries led to a burgeoning population that drink up most of the river's day-to-day flows, and quickly also pumped dry much of the groundwater below the river. The river began to run dry for much of the year, and forgetting the flood risk, people and industry began to settle in the floodplain. Sand and gravel were extracted from the river bed, which became a dumping ground for trash (Gumprecht 2005).

When the flood waters returned, perhaps most notably in a massive flood in March 1938, catastrophic damage to people and property ensued. This led to the US federal government taking over responsibility for managing the river. The US Army Corp of Engineers redesigned the river with a singular focus on flood control. Most of the river was lined with concrete, with a goal of speedily moving flood water out of the city and to the sea. As surface water that flowed off LA's streets was so contaminated, the goal of city policy came to be limiting stormwater flows into the river during low-flow periods to prevent this pollution from leaking into groundwater aquifers. The strategy, in essence, was to disconnect the river from the city as much as possible during low-flow periods, and then to dump flood water into the river when necessary (Gumprecht 2005).

To many in LA, the river became an invisible place. Hence Rorabach's tongue-in-cheek exploration of the river, like he was an adventurer charting an unknown watercourse. He described what he found as the "traditional LA marshland of cattails and shopping carts" (Gandy 2014).

To those who lived adjacent to the river, it was more complex (Fig. 7.1). The river was a place for kids to go to escape the violence in their neighborhood and just play and explore. But it also sometimes became a place where the homeless and those addicted to drugs congregated. Ecologists can criticize how it is bad for nature when a river loses all ecological

Fig. 7.1 The Los Angeles River, as seen from the 4th Street Bridge facing north. (Photo credit: Adan Garcia (CC BY 2.0))

function. But it is also profoundly damaging for humanity when a river becomes a gray place, when a concrete-lined canal becomes the closest park for children to play in.

In the 1980s, however, a movement began to revitalize the river. Friends of the LA River and other community groups began to form. The first Great LA River Cleanup was held in 1989, and a similar gathering to remove trash has been held every year since. Citizen demand for a cleaner and greener river pushed the city and the US Army Corps of Engineers to begin planning for a different way to manage the river (Perini and Sabbion 2016).

In 2007, a greener vision was articulated in the city's LA River Revitalization Master Plan. The plan strives to create a green strip through the city along the river corridor, with bikeways, pedestrian paths, and public open spaces (City of Los Angeles 2007). Hydrologically, the goal was to enhance the quality of water flowing into the river using stormwater green infrastructure, which would also serve to slow the flow of rainwater into the river and enhance groundwater recharge. The plan calls out two main interventions. One is use of stormwater green infrastructure to clean and delay stormwater outflows, as described earlier. An early example of this was in 2008, when the Dominquez Gap Wetlands project began. Another was the creation of new riverfront and linear parks. Progress has been steady but slow toward the LA River Revitalization Plan's vision. In June 2017 California voters approved a $4 billion bond for the protection of open space for recreation or water quality, and it is hoped that some portion of this money will be able to accelerate change along the LA River (Perini and Sabbion 2016).

Los Angeles had already destroyed its river ecosystem by the 1980s and faced a choice (Fig. 7.2). They could continue ignoring the river, letting it be a gray, bleak place. Or they could try to restore it, for the benefit of both people and nature. They have chosen the latter path. The river, of course, will never be fully restored to a fully natural state, and it would have been far cheaper for the city to have planned initially to protect the river, rather than to have to do extensive, expensive restoration. Nevertheless, even committing to restoration takes courage to believe that something beautiful and natural can return to such a gray space.

The Stakes

We live in a world filled with many places like the LA River, soulless spaces where concrete and asphalt have replaced leaves and branches. One study measured the average forest cover in neighborhoods in 245 cities around the world (McDonald et al. 2018). Only 13% of neighborhoods had forest cover at levels that have been shown to be supportive of greater mental health. While there are many more ways to measure nature exposure in cities (see Chap. 4), the study suggests that far from being unusual, what happened to the LA River is quite common. Most of our urban spaces are

Fig. 7.2 A different view of the Los Angeles River, in one of the few stretches that was not channelized. The resurgence of interest in the river has led to increased demand for recreation along the river. (Photo Credit: Los Angeles District (CC BY-ND 2.0))

mostly gray and lack enough nature to overcome the urban mental health penalty.

We are currently on track to build even more of these lifeless places. Cities in many countries around the world have been losing tree cover. In the United States, urban tree cover between 2009 and 2014 declined by 0.7%, an estimated 36 million trees (Nowak and Greenfield 2018). Global trends are more uncertain, although there seems to be different trends by region and country (McDonald et al. 2018). Despite the increasing evidence from the scientific community of how important nature is for health, *Homo sapiens* is still building an urban world mostly lacking in nature.

If we continue building cities like this, we risk condemning billions of people to suffer from the urban mental health penalty. Life in cities without nature is associated with increased incidence of depression, schizophrenia, and other ills. Beyond that, what is at stake as we build the urban neighborhoods of the future is our happiness and joy in our new urban world. Without nature in our cities, they will fail to be vibrant, thriving places to live.

The Choice

Los Angeles' decision is one that will face thousands of cities around the world. We can continue our current path, paving over and replacing nature with an entirely gray world. Or we can follow another path toward a more natural urban future. We could protect key natural habitat where it exists, siting urban growth for other locations. We could restore or create natural features where nature has been destroyed. There are 1860 cities globally with more than 300,000 people living in them (UNPD 2018), and many thousands of smaller urban settlements. All these urban settlements will have to choose which of these paths to follow.

There is an irony that in the urban century, just as the vast majority of *Homo sapiens* move into cities, nature is becoming more necessary than ever. The urban pattern of settlement brings many advantages to humanity, among them economic development, increased creativity and innovation, and efficient resource utilization. But cities are also quite unnatural, creating an urban mental health penalty. Nature in and near cities can be a way to have our cake and eat it too, to have the benefits of our urban world while still having sane, thriving, beautiful places to live.

Some readers might rightly ask whether it is unfair to ascribe to humanity at large a choice about the future of urban nature. Real decisions in cities are a complex function of the interplay of several different powers. Government politicians, civil engineers, agency bureaucrats, captains of industry, and community leaders all shape the decisions that are made. There are deep inequities of power in cities, and we should acknowledge that in many cities some people have little real power to affect this "choice." This is particularly true for the roughly one billion who live in informal urban settlements (UN Habitat 2016), who are often intentionally excluded from governmental decision-making. Similarly, for the billions who live in countries with non-democratic governments, it can be hard to influence government decisions.

The reality of urban planning decisions may be of course less binary than we have presented here, for there is a continuum of degrees of nature exposure in cities. But we urge that cities do not lose sight of the fact that they must decide how to incorporate nature into the future. A more verdant urban future is possible, if we have the collective will to choose that path.

Challenges in Making the Biophilic Choice

We have argued in this book that biophilic design and planning is one of the most helpful frameworks for protecting and restoring the nature we have in cities, while fostering new connections with the natural world. However, cities making the biophilic choice will not find it easy and may confront some serious obstacles along the way. A few of these are briefly mentioned in the following section. We believe there is a need to confront them and address them for a city to be successful in achieving its biophilic vision.

Concerns About the Cost of Biophilic Design and Planning

Most of the biophilic design and planning ideas discussed in this book are highly cost-effective: a green rooftop, for instance, while adding some additional upfront cost to the construction of the building, can often pay for itself (by, among other things, extending the longevity of the underlying roof). Equally true, the costs of planting a tree (or better yet protecting and saving existing mature trees) is more than outweighed by the public (and private) economic benefits and value produced.

The problem is often how to cover the short-term costs of such measures. As discussed in Chaps. 5 and 6, there are city programs that provide incentives to offset this short-term cost, such as a density bonus for the installation of extensive green roof, or a reduction in stormwater fees for tree planting and reduction in impervious surfaces.

The Relevance of Biophilic Design and Planning to the Global South

A common critique of the vision of green cities as advocated for in American and Northern European contexts is whether and in what ways the model can apply to the growing cities in the Global South. In such cities, it is argued, the urban planning challenges are more basic and must necessarily focus on poverty-reduction and provision of basic goods such as clean air, water, and food. Here relatively large percentages of population live in informal housing and settlements, often in more dangerous and marginalized parts of cities. Further, it is argued that cities of the Global South will have fewer resources to spend on parks, greenspaces, and trail systems, which may be viewed as luxuries. We recognize these are legitimate concerns and that additional work must be done to find ways to apply the biophilic urban vision to cities in the Global South. However, we believe that the biophilic vision and the tools for its implementation are equally if not more applicable to cities in the Global South. For example, tree planting is proving to be a highly cost-effective response to air pollution, and designs that facilitate growing food in urban neighborhoods and other investments in nature will help residents of these cities become more resilient. We now have a growing set of case studies and examples of how nature-based solutions can address a variety of tough problems in cities with fewer resources (e.g., Beatley 2017).

Avoiding Eco-gentrification

A connected set of issues arises from the unintended consequences of some urban biophilic projects, such as the construction of a new park or other green enhancement. Some projects like the High Line in New York have the undesired result of raising the cost of housing and displacing residents. As a form of gentrification, these effects are sometimes referred to as "eco-gentrification," and are a troubling unintended consequence of otherwise exemplary projects like the High Line. While it is not clear if most urban greening projects have significant impacts on the cost of housing, gentrification around projects like the High Line suggests that sometimes they can. There is a need to develop new policies in cities to ensure that housing remains affordable and that the benefits of urban greening projects are shared equitably. New models of equitable development are also needed and are under development. The 11th Street Bridge Park in Washington, DC, is one such emerging positive example. It will become an important biophilic park in the city (a pedestrian bridge and park that will span the Anacostia River), and from the beginning, its designers have engaged the local neighborhood. While crafting the park's equitable development plan, several actions have been taken to ensure that affordable housing remains in the neighborhood, including the formation of a community land trust, and that residents will be able to benefit from the jobs created from the project. Any vision of biophilic cities must be cognizant of these kinds of unintended consequences and work to ensure that benefits from investments in local nature will be shared fairly.

Smart Cities vs./and Biophilic Cities

Few visions of future cities have captured the attention and fascination of urbanists as quickly as has the "smart city" agenda. There is little question that the deployment of new communication and sensing technologies opens new possibilities in the way cities will look, feel, and function. Does a smart city compete in some serious way with our vision of biophilic cities? Could the visions be combined or productively integrated? Possibly yes. It is important to say first that the vision of smart cities that has been advanced in the popular media, with its renderings of technological connectedness, seems rarely to include nature or biodiversity. To us, the smart city vision often suffers from the same natureless view that we find in much mainstream architecture and urban design. But there is potential, we think, to utilize these emerging technologies—the connected city of sensors—on behalf on the vision of biophilic cities. Already we have examples of green rooftops sensors that send messages when vegetation needs watering, and similar approaches could be taken to trees and urban gardens in ways that might prevent plant and tree mortality. Nadine Galle et al. (2019) have argued for what they call "The Internet of Nature," that carries this idea forward in interesting ways, imagining how a variety of new technologies including drones, robotics, biosensors, augmented reality, and wearable technologies could enhance our connections to nature and help to grow, protect, and preserve the nature around us in cities.

Future of Urban Nature-Hybrids

We are now in an interesting period when the very idea of "nature" is changing, and where we see the emergence of "hybrid" forms of "nature-structures." The Supertrees of Singapore represent one example of this—human-designed, metal structures that are not natural in the same way a tree is. However, they create spaces for thousands of hanging plants, and in this way can support the existence of some natural features in cities. Consider also some of the new indoor biophilic spaces, such as Jewel, a seven-story indoor waterfall and rainforest located at Singapore's Changi Airport. Jewel is large enough to have hiking trails and houses some 1400 trees. These spaces are, of course, highly designed, yet do provide considerable human benefits and an important element of connection to nature in dense urban settings. Questions will remain about their "authenticity" as real nature spaces and whether the resources used to create these spaces

($1.7 billion in the case of Jewel!) could be better spent in protecting and conserving wetlands, forests, or other actual living natural systems in the community.

The Indoor-Outdoor Conundrum

The reality of modern life is that much of our time is spent indoors, away from sun, trees, birds, and breezes. There is without question a human need for shelter, from weather and climate, but we have constructed modern cities that assume that inside is the best place to be. It is a conundrum and a challenge. There are a couple approaches cities can consider, and perhaps a blending of all of them is appropriate. The first is the idea of working to design cities that will reward being outside—cities that reflect the immersive forms of nature we have discussed in this book. Biophilic cities like Oslo, San Francisco, and Singapore often invest in urban trail networks that make it safe, convenient, and fun to spend time outside. In higher-latitude cities, getting people outdoors can require special strategies. Edmonton, Canada, for instance, has developed winter strategies to entice residents outside, creating destinations and activities to lure residents out of doors when cold temperatures would otherwise keep residents away.

Second, there is a need to adapt building design to the unique climate and climatic opportunities that will exist in different parts of the world, often building on to long cultural traditions and patterns of living. Warmer Mediterranean climates such as in southern Europe, California, Australia, or South Africa provide the chance to blend the indoor-outdoor bifurcation. Australians capture the outdoor spaces around and behind a home, extending the living quarters to include such outdoor rooms. In colder climates in many mid- and upper-latitude cities, the strategy is to bring more nature into the interior spaces of homes and offices. There is a growing trend—and an increasingly rich pallet of design ideas and commercial products to support it—to create interior water walls or compact growing systems for growing vegetables.

Cities Working on Behalf of Global Nature

Yet another significant open question is whether and in what ways biophilic cities can not only create the conditions for conserving nature locally but also help to address nature and biodiversity conservation globally. This question points out the paradox of affluent cities such as New York or Singapore taking effective actions to create healthy biophilic conditions locally but, due to their outsized ecological footprints, causing loss or destruction of biodiversity elsewhere (McDonald et al. 2020). In response we believe that the practice of biophilic cities must include actions and leadership beyond local boundaries. This could take many different forms, including city-to-city conservation agreements, efforts to financially underwrite conservation elsewhere, or efforts to reduce the extent of unsustainable resource flows (for instance, the carbon footprint associated with a city's consumption of food). Biophilic cities, moreover, should consider flexing their political and economic muscles on the world stage in support of bold conservation ideas such as protecting Half-Earth (Wilson 2016; Dowton 2017). If biophilic cities are truly "cities that love nature," that love should extend to more distant nature as well as the local nature directly enjoyed by a city's residents.

What You Can Do

We believe that there is a role that everyone can play in choosing a more verdant urban future. Everyone reading this book can become advocates in their own way. You can visit natural areas in your city and begin to learn more about their ecological health and their history. You can talk to your friends and family about what nearby nature means to you. Where possible, you can reach out to government officials and let them know what kind of urban world you would like us to pass on to future generations.

Community organizations will play a key role in choosing our urban future. In Los Angeles, groups like Friends of the LA River were key to pushing for a cleaner and greener river. In many cities around the world, there are similar organizations. Moreover, networks of cities like the Biophilic Cities Network and ICLEI serve to unite community organizers across cities. We hope this book inspires many of you to become more active in community organizations or start your own.

One intended audience of this book is urban planners and landscape architects. Planning practitioners have a central role to play in designing a more natural urban future. Without the planners who created the 2007 LA River Restoration Plan, much of the current progress along the river would not have occurred. There are emerging best practices for how to incorporate nature into urban planning. The Biophilic Cities Network website (www.biophiliccities.org/) contains case studies of biophilic design, while the Greenprint Resource Hub (www.greenprinthub.org/) lists guidelines for incorporating biodiversity and ecosystem services into urban planning.

Finally, we hope this book inspires government officials. Political leadership from the city of Los Angeles has been key to changing the river for the better. Whether at the city, subnational, or national level, leadership by government officials is key to choosing a more verdant urban future.

The Urgency of Now

It is important to realize that this choice between a gray and a green urban future is not some hypothetical future decision facing cities but is a decision that is occurring right now. Every day, countless cities make decisions, large and small, about what their urban form will be in the future. We are in the fastest period of urban growth in human history. Every six weeks, *Homo sapiens* builds urban area that is equivalent in population to London (UNPD 2018). Every week that goes by, therefore, without further global recognition of the importance of nature in cities is a week of opportunities lost, of bad decisions made.

Moreover, the decisions we make now will also have implications for generations to come. Urban form, once built, persists for decades or centuries, only slowly changing. Many European cities, for instance, still retain a ring road and a dense city center, a vestige of medieval city walls pulled down centuries ago. More proximate in time, the spat of highways built after World War II in many countries have created a large sprawl of low-density neighborhoods that persists even today. Similarly, the legacy of our choices about urban nature today will affect access to urban nature for many future decades.

Parting Thoughts

We will get, ultimately, the urban world we choose. In a certain sense, we will get the urban world we deserve. Even ignorance of the value of urban nature is a sort of choice, a choice to continue on the status quo path. In many ways, the future of nature in cities will be determined by how much people in cities demand nature. The human psyche is shaped by the presence of nature, but the human psyche will also determine how much urban nature there is.

The future of urban world is contested along multiple different dimensions, of which the question of urban nature is only one. There are multiple visions of what our new urban world could be, all fighting for dominance in the marketplace of ideas. We believe that fighting for urban nature is fighting in some ways for the preservation of the spirit of humankind. If biophilia is correct, and we all have an innate love of nature, then fighting for a more verdant future is in some ways fighting for some deep-rooted part of ourselves.

We have described the dramatic tension we see between the benefits of urban proximity and its concomitant negative urban mental health penalty. We have argued in this book that nature in and near cities is an essential part of reducing this tension. We wrote this book to inspire action, by all of you reading this, in the hopes you will join us in fighting for this more natural urban future.

References

Beatley, Timothy. 2017. *Handbook of biophilic city planning & design*. Washington, DC: Island Press.

City of Los Angeles. 2007. *Los Angeles River revitalization master plan*. Los Angeles: Bureau of Engineering. http://boe.lacity.org/lariverrmp/.

Dowton, P. 2017. Half-Earth cities. The Nature of Cities. https://www.thenatureofcities.com/2017/12/26/half-earth-cities/.

Galle, N., S. Nitoslawski, and F. Pilla. 2019. The internet of nature: How taking nature online can shape urban ecosystems. *The Anthropocene Review* 6 (3): 279–287.

Gandy, M. 2014. Tracing the LA River. In *The fabric of space: Water, modernity, and the urban imagination*, ed. M. Gandy. Cambridge, MA: MIT Press.

Gumprecht, B. 2005. Who killed the L.A. River? In *Land of sunshine: An environmental history of metropolitan Los Angeles*, ed. W. Deverell and G. Hise. Pittsburgh: University of Pittsburgh Press.

McDonald, R.I., T. Beatley, and T. Elmqvist. 2018. The green soul of the concrete jungle: The urban century, the urban psychological penalty, and the role of nature. *Sustainable Earth* 1 (1): 3.

McDonald, R.I., et al. 2020. Research gaps in knowledge of the impact of urban growth on biodiversity. *Nature Sustainability* 3: 16–24.

Nowak, David J., and Eric J. Greenfield. 2018. Declining urban and community tree cover in the United States. *Urban Forestry & Urban Greening* 32: 32–55.

Perini, K., and P. Sabbion. 2016. *Urban sustainability and river restoration: Green and blue infrastructure*. John Wiley & Sons.

Roraback, D. 1985. The L.A. River practices own trickle-down theory. *Los Angeles Times.*

UN Habitat. 2016. *Slum almanac 2015–2016: Tracking improvement in the lives of slum dwellers.* Nairobi: Participatory Slum Upgrading Programme, UN Habitat.

UNPD. 2018. *World urbanization prospects: The 2018 revision.* New York: United Nations Population Division.

Wilson, E.O. 2016. *Half-earth: Our planet's fight for life.* New York: WW Norton & Company.

Index

R. McDonald, T. Beatley, *Biophilic Cities for an Urban Century*,
https://doi.org/10.1007/978-3-030-51665-9

manufacturer's authorised representative in the EU is Springer
Customer Service Centre GmbH, Europaplatz 3, 69115 Heidelberg,
. If you have any concerns regarding our products, please
ductSafety@springernature.com

und by CPI Group (UK) Ltd, Croydon, CR0 4YY
28/09/2026
02224088-0002